T0168878

UNDER DESERT SKIES

MELISSA L. SEVIGNY

UNDER DESERT SKIES

How Tucson Mapped the Way to the Moon and Planets

SENTINEL
PEAK
TUCSON

SENTINEL PEAK
An imprint of The University of Arizona Press
www.uapress.arizona.edu

Printed in the United States of America
21 20 19 18 17 16   6 5 4 3 2 1

ISBN-13: 978-1-941451-04-5 (paper)

Cover design by Leigh McDonald
Cover photo courtesy of William K. Hartmann and Ewen Whitaker, Hon. D.Sc.

The royalties for this book have been donated to support space exploration and science education efforts at the Lunar and Planetary Laboratory.

Publication of this book is made possible in part by funding from the University of Arizona's College of Science and Lunar and Planetary Laboratory.

Library of Congress Cataloging-in-Publication Data
Sevigny, Melissa L., 1986– author.
  Under desert skies : how Tucson mapped the way to the Moon and planets / Melissa L. Sevigny.
    pages cm
  Includes bibliographical references and index.
  ISBN 978-1-941451-04-5 (pbk. : alk. paper)
    1. University of Arizona. Lunar and Planetary Laboratory—History. 2. Planetary science—History.
I. Title.
  QB501.S45 2016
  559.9—dc23
                        2015029710

♾ This paper meets the requirements of ANSI/NISO z39.48–1992 (Permanence of Paper).

In memory of Mike Drake
(1946–2011)
scientist, mentor, friend

CONTENTS

ILLUSTRATIONS

UNDER DESERT SKIES

PROLOGUE

WHEN I FIRST came to work for Michael Drake, then director of the University of Arizona's Lunar and Planetary Laboratory (LPL) in Tucson, he described my task in a single line: "Capture the history of LPL from the old timers while they're still alive and turn that into a transcript, if not a narrative."

The project took the better part of four years. I left LPL with more than fifty interviews recorded and transcribed, and I still felt the work was only half-done. Whenever I met with scientists to document their experiences in Tucson, from its first forays into the Space Age in the 1960s to its present prominence in solar system exploration, I looked forward to asking one particular question. I asked them to tell me about the first moment they saw a new world revealed.

Some told stories about standing in the chill emptiness of a telescope dome, gazing at a distant point of light and watching something extraordinary unroll in a printout of numbers. Others described a photograph snapped by a spacecraft and transmitted to Earth at a moment when no one else stood in the command room to receive it. These "first sights" ranged from the Moon to Mars, from asteroids to the moons of asteroids, from the ultraviolet drama of photons skipping off Earth's atmosphere to the invisible bubble of the heliosphere, which marks the outermost edge of our Sun's influence.

In those stories, I knew that I had found the heart of what it meant to study planetary science: an inexhaustible sense of wonder.

My own "first sight" took place on May 26, 2008, in the engineering test bed of the Science Operations Center (today called the Michael J. Drake Building). The Phoenix Mars Scout Mission, named for the mythological bird because it rose from the ashes of a previous failed mission, had landed safely on Mars the evening before. I headed home late that night exhilarated and exhausted. I paused in the parking lot to look up in bewildered joy at the red speck resting easy in the western sky, above the sleeping contours of the Tucson Mountains. So much vast distance between the two planets, crossed, if not in an instant, then at least in the 15.3 minutes it took the Phoenix lander to transmit its touchdown signal to Earth on Landing Day.

The slapdash cubicles belonging to the education and public outreach team, where I worked for the length of ground operations, were tucked into a corner of the payload interoperability test bed, a warehouse-sized room that echoed with every step. A double of the Phoenix Lander rested on a platform nearby, surrounded by fake Martian rocks—a practice ground that engineers would rearrange into a close facsimile of the landing site.

Black paper over the windows blocked out the fierce desert sunlight to help the scientists adjust to working on "Mars time," which meant adding an additional thirty-seven minutes to each day. They would wake and sleep with the solar-powered lander. It felt almost like going to work on Mars itself, an impression that strengthened as Phoenix relayed its daily stream of data.

The Phoenix team had embarked on this latest mission at an unprecedented time in human history. No fewer than three orbiters circled above Mars' thin pink atmosphere, and the rovers Spirit and Opportunity still explored the equator. Almost all of those missions carried instruments or drew on expertise that had originated at LPL. I had only recently embarked on Mike Drake's quest to learn and record what brought Tucson to this prominence—an odd mix of geography and personality helped by the deep clarity of the dark desert skies.

One such instrument, HiRISE (High Resolution Imaging Science Experiment), the camera aboard Mars Reconnaissance Orbiter, had photographed Phoenix's landing site in Vastitas Borealis for the last several years, recording seasonal frost recede from the surface to reveal a puzzle-piece landscape. Just eleven hours after touchdown, the HiRISE team captured a photograph of the lander from their headquarters in another campus building. Bright blue in the color-filtered image, Phoenix bravely spread its solar panels to the cold. From the vertical distance of roughly two hundred miles, the lander looked like a molecule of the water we hoped to find, three linked circles in a sea of red.

Many images would arrive from Phoenix over the next five months, some black-and-white and grainy, others compiled into magnificent panoramas. But that speck of blue stuck in my mind. To me, it meant the fulfillment of my childhood dream to travel to another planet—if not with my own two feet, then at least through the firmly planted tripod of a spacecraft.

For LPL, the portrait of the Phoenix lander symbolized three triumphs. First, its artistry was a tribute to the scientists on the HiRISE team, who had also captured with incredible precision a picture of Phoenix plummeting to the ground, parachute streaming—the first image ever taken of a spacecraft's descent to another planet.

Second, the flat red terrain pictured beneath the winged circle of the lander belonged to the success of yet another LPL instrument, the gamma ray spectrometer on the Mars Odyssey orbiter. Six years earlier the spectrometer had detected vast quantities of hydrogen lying beneath the snakeskin surface of the poles, the H in H_2O. The discovery was compelling evidence that water still existed on Mars, locked away in ice.

And then Phoenix itself. Designed to dig into the Martian surface with a robotic arm in search of water and organic molecules, Phoenix represented the first time a public university controlled ground operations of a NASA mission. When the spacecraft confirmed the presence of frozen water a scant two inches beneath the dust, one LPL scientist would say with a hint of exasperated humor, "We discovered water on Mars—*again*." In a single snapshot, three remarkable successes conspired to meet at 68 degrees north and 234 degrees east, on a world that couldn't be more different from Tucson's blazing heat.

Many stories can be told about LPL's journey from a small research lab to a leader in planetary science, and how this corner of Arizona became Earth's ambassador to the solar system. This book does not capture them all. But in a small way it fulfills another dream—Mike Drake's hope of crafting a narrative that would preserve the triumphs, challenges, and extraordinary journeys of those who soon would be gone. In the midst of recording the memories of LPL scientists, I felt fortunate to have my own story of a "first sight" to tell. Nearly five decades after its founding, LPL's legacy alighted on the northern plains of Mars. The Phoenix lander blinked its message to Earth: at last, we have arrived.

PART ONE

ARRIVAL

I N LATE AUGUST 1955, at the Ninth General Assembly of the International Astronomical Union (IAU) in Dublin, Ireland, the Dutch-born astronomer Gerard P. Kuiper circulated an unusual memo: would anyone help him create an atlas of the Moon?

Out of some five hundred astronomers present, only Ewen Whitaker replied. Few scientists in those days studied the Moon. Telescopes had long ago revealed its barrenness, and most astronomers dismissed the Earth's nearest neighbor in favor of dim galaxies and distant star clusters. When Whitaker joined the Lunar Section of the British Astronomical Association, a society for amateurs, he was dismayed to discover a hodgepodge collection of hand-drawn lunar maps. After two years of exchanging letters and ideas, Whitaker accepted Kuiper's invitation to fly out to Yerkes Observatory in Williams Bay, Wisconsin, and spend a month printing lunar photographs in the darkroom.

At the London airport, Whitaker snagged a copy of the evening newspaper and took it on the overnight flight. The date was October 5, 1957. Bold-lettered headlines declared the Soviet Union's successful launch of Sputnik 1, the first human-made object to reach outer space. In the morning, Kuiper came to meet Whitaker in Chicago, and Whitaker handed him the paper.

"Oh, I haven't even heard about this," Kuiper said. "This is big news."

Determined, energetic, and headstrong, Gerard Peter Kuiper had made a name for himself in the scientific world. Like Whitaker, he had dedicated himself to the study of the solar system, an uncommon choice at that time. Sputnik's glimmering trail across the autumn sky *was* big news in ways that the two astronomers could not foresee. In the frenzy of the Space Race, the long-disregarded Moon would suddenly become an object of interest.

BEGINNING

Kuiper's Lunar Project flourished at Yerkes Observatory, managed by the University of Chicago. In 1958 Whitaker made the fateful decision to move his young family from England to America to work on the project full-time. Kuiper also hired David ("Dai") Arthur, an outspoken Welshman who had worked professionally in Great Britain as a cartographer and found an outlet for his love of astronomy in the Lunar Section of the British Astronomical Association.

The two Brits had settled on the stately grounds surrounding Yerkes Observatory by October 1958. The same month, a new agency created by President Dwight D. Eisenhower began its operations: the National Aeronautics and Space Administration (NASA), a civilian organization that inherited space-centered research labs from around the country.

Though Arthur and Whitaker were amateur astronomers, Kuiper cared more about their unparalleled skill as observers than their lack of academic degrees. The three scientists published the first edition of the *Photographic Lunar Atlas*, a twenty-two-pound box of loose-leaf sheets, in 1960. They immediately set to work on a supplement that would add precise coordinate grids, which would be published as the *Orthographic Atlas of the Moon*, editions A and B.

By this time Kuiper's group had clearly outstayed their welcome at Yerkes. Twice appointed the observatory's director, Kuiper's single-minded devotion to the Moon and planets to the exclusion of all else affronted the gifted scientists around him. Discontented colleagues and perpetually cloudy skies prompted Kuiper to look for a different university to continue his ambitious project. He chose Southern Arizona for the same reason that most people do: the weather. Clear skies and low humidity ensured that there would be good "seeing" conditions for astronomers nearly year-round. He envisioned the peaks of the Catalina Mountains adorned with the silver silhouettes of telescope domes.

The president of the University of Arizona, Richard Harvill, welcomed Kuiper's group. Harvill saw it as yet another outstanding research organization to bolster the university's reputation. Steward Observatory, then directed by Edwin Carpenter, was fast becoming a leader in astronomical research, and Kuiper's colleague Aden B. Meinel had recently spearheaded the effort to establish nearby Kitt Peak as the site of a new telescope observatory.

But not everyone heralded Kuiper's arrival with enthusiasm. Planetary studies had fallen out of fashion since the late nineteenth century, when the marriage of cameras and telescopes made it possible to study faint galaxies, star clusters, and nebulae. Astronomers had long cursed the rise of the incorrigibly bright Moon. The aversion arose from a practical matter. The appearance of the Moon in the night sky signaled the end of an observer's precious "dark time." It seemed strange to Tucson astronomers that this new group wanted to devote valuable resources to studying such a predictable and uninteresting object.

Tension between the competing interests of Kuiper and Carpenter posed a problem for President Harvill: where could he house the new planetary sciences lab? A. Richard Kassander Jr., director of the Institute of Atmospheric Physics, made a fortuitous offer. The newcomers could have a small section of the new Physics, Mathematics and Meteorology (PMM) Building, scheduled for completion in the summer of 1960.

Whitaker and his family arrived in Tucson in the sweltering heat of August, in a Rambler American station wagon jammed with boxes and furniture carefully arranged to keep the kids separated during the long drive. Kuiper followed later that month, along with six of his Yerkes staff: Dai Arthur, Barbara Middlehurst, Carl Huzzen, Ruth Horvath, and Elliott and Gail Moore. Two graduate student assistants, Toby Owen and Ann Geoffrion, joined the group, bringing the new laboratory's personnel up to ten. With superlative self-confidence, Kuiper purchased a house in Tucson before receiving a formal job offer from the university.

"We started up in very humble surroundings," Whitaker said. Kassander's staff moved into their new building from temporary quarters in Quonset huts, half-cylinder structures with curved roofs left over from World War II. Kuiper's group claimed a house-sized corner at the west end of the top floor of the PMM Building, today called Physics and Atmospheric Sciences. They retained one of the Quonset huts for extra space. Kuiper called his new group the "Lunar and Planetary Laboratory of the Institute of Atmospheric Physics."

"We moved into this place from the Quonset huts, set up our darkrooms and got on with the work of the lunar atlas," Whitaker said. "That's how it all started."

STUDENTS

Astronomers had little liking for the new field that would later become planetary science, the disfavored stepchild of astronomy. Telescope time could not be wasted on such nearby objects as planets. Percival Lowell's imaginative descriptions of a Mars webbed in a complex canal system in the early twentieth century did nothing to improve matters. No one wanted to study the Earth's near neighbors: they were the stuff of science fiction, not science.

President John F. Kennedy's announcement in 1961—that an American would stand on the Moon before the end of the decade—sent a shockwave through the scientific world. At the time, scientists knew very little about the composition of the Moon or the nature of the dark areas that early astronomers had sketched in their notebooks and called *maria* (seas). How could America possibly send humans there with so little knowledge?

Only three notable scientists seriously studied the solar system in the United States: Fred Whipple at Harvard, Gerard Kuiper in Arizona, and Harold Urey in California. Kuiper had formed a lasting friendship with Whipple, matched by an equally lasting enmity with Urey, who had engaged him in a vitriolic debate over the origin of the Moon. For students wishing to study planetary astronomy—the field of planetary science did not yet exist—Urey's school in San Diego and Kuiper's competing lab in Tucson were among the few places they could go.

In 1961, four newcomers arrived at the Lunar Lab: graduate students Alan Binder, Dale Cruikshank, and William Hartmann, and undergraduate Charles Wood. Binder and Cruikshank had both worked with Kuiper at Yerkes Observatory and followed him to Tucson after finishing their bachelor's degrees. Hartmann arrived from Pennsylvania State University, where he had studied astronomy. Along with Toby Owen and a handful of other student assistants, they were the first informal graduate students of the Lunar Lab, which at the time did not offer a degree.

These were children of the Sputnik era, their imaginations fired by the same rockets that carried the first satellites into space. Chuck Wood had spent his childhood Tuesdays listening to the exploits of Tom Corbett, Space Cadet, on the radio. In college Bill Hartmann pored over the pages of *Collier's Weekly* to find articles by Wernher von Braun, illustrated with the otherworldly paintings of Chesley Bonestell. Only a decade before, von Braun predicted that rockets might one day fly to the Moon. At last it became possible to live that dream.

"Kids growing up in the fifties were immersed in the idea that we were going to travel into space—that science fiction was going to become real," Wood said. As an undergraduate in astronomy at the University of Arizona, he ran across a newspaper clipping announcing Kuiper's arrival in Tucson. Fascinated, he spent six months writing letters and making phone calls until finally obtaining a job with the new lunar group.

"A couple of hours a day after classes, I would use rulers and measure craters," Wood said. "It was before computers were widely available. We used these old-fashioned calculating machines. You'd key in the numbers, like a cash register, and then you'd pull the crank multiple times and it would *chug, chug, chug,* and finally multiply two numbers together. I did that for four years. I worked my way through school as an undergraduate by measuring craters on the Moon."

Kuiper split his new assistants between the lunar cartography program and the study of planetary atmospheres. The students crowded into a Quonset hut called Temporary Building Number Six (T6), working in the coolness of the development lab. "There used to be jokes about Kuiper flying into a tizzy over something and saying, 'Call T6, call T6,' because a bunch of us graduate students over there were about to be chewed out," Hartmann said. They soon discovered that Kuiper had little interest in teaching. He considered graduate students simply a necessary support for his lunar project. They would learn on the job, or in other departments.

With no planetary science degree available—not just at the University of Arizona but anywhere in the country—the Department of Astronomy seemed the natural choice. Edwin Carpenter had trouble accommodating the new arrivals. His department offered foundation courses in physics and mathematics to prepare students for the complexities of stellar interiors and galactic structures. As demand grew, the department expanded its curriculum to include classes on the solar system, often taught by professors who held joint appointments with the Lunar Lab.

Many of Kuiper's graduate assistants chose to pursue these hybrid astronomy degrees, including Toby Owen. He had fallen in love with the night sky as a kid in Santa Fe. Owen met Kuiper at the University of Chicago and followed the esteemed astronomer to Tucson after earning a master's degree in physics.

"Learning the constellations and watching the planets move among the stars was a pleasure that never left me," he said.

Kuiper taught Owen spectroscopy, the study of how light breaks into spectral lines. As light interacts with matter, its spectrum is marred with emission or absorption lines, each one a unique "fingerprint" that allows scientists to infer the molecules in a planet's atmosphere. Kuiper took his state-of-the-art spectrometer on observing runs to Kitt Peak, driving a pickup truck up the steep, dusty road with students "hanging on for dear life" in the back, as Cruikshank recalled.

But the art of interpreting a ray of light—the heart of astronomy—was not sufficient for the new field that Kuiper wanted to forge. Therefore a few graduate assistants found help elsewhere. Spencer Titley, a professor in the Department of Geosciences, took Binder, Cruikshank, and Hartmann under his wing.

Titley specialized in ore deposits, but his heart was in space. Aspiring to join the Astronaut Corps, he wrote letters to NASA campaigning for more scientific involvement in Project Gemini, which would launch manned spaceflights to develop space travel techniques in support of the Apollo program. Titley also collaborated with Eugene Shoemaker's new Astrogeology Research Program. Shoemaker had created the institution as a branch of the United States Geological Survey (USGS) in 1960 to pursue lunar mapping. Three years later the Astrogeology Research Program moved to Flagstaff, Arizona, where nearby Meteor Crater made an ideal training ground for astronauts headed for the Moon.

"There was a coincidence of these three fellows coming in and my involvement with the Survey on this new, exciting thing with lunar mapping," Titley said. In the late sixties Binder and Cruikshank received PhDs in the Department of Geosciences, while Hartmann earned a master's in geology and a PhD in astronomy, all writing dissertations focused on the Moon. A few years behind them, Wood completed his master's thesis on the steam-exploded craters in the Pinacate region of Sonora, Mexico, in the early seventies.

"I hand-tooled them, in a sense," Titley said. "We had special courses, and I tried to take physicists and astronomers and turn them into geologists—fairly successfully, I think, because they were bright people. That was simply how the program worked."

Geology courses turned out to perfectly complement the students' arduous assistantships in the Lunar Lab. Under Titley's guidance, the graduates charted lunar quadrangles on newly printed maps and took trips to the Pinacate lava fields along the Mexico border. Among cinder cones and calderas, the students

FIGURE 1. Allen Thompson, Steve Larson, Godfrey Sill, and Dale Cruikshank explore the Pinacates in Sonora, Mexico, December 1967.

COURTESY OF D. C. CRUIKSHANK

learned a geologist's perspective of the lunar surface. They imagined hiking it, measuring it, and chipping samples from the stones.

"Kuiper understood that the study of planets was not astronomy anymore," Binder said. "Clearly one had to understand geology and geophysics and so forth. So it moved in this direction, but in the beginning it was very confusing as to what we were supposed to do and how we were supposed to get educated."

The students forged close friendships, camping in Mexico on weekends or escaping the sweltering summers in movie theaters. "I remember when we saw *West Side Story*," Wood said, "and we all came out walking in a line and snapping our fingers like we were the Jets." The camaraderie of the small group, which set a precedent for the Lunar Lab's students that would continue for decades to come, rose from their common interest. The solar system had just begun to come into its own as a respectable subject, and the young students gambled their careers on it.

EXPANDING

At the time, a low basalt wall contained the university's buildings, built with dark, jagged stones quarried from nearby Sentinel Peak. The wall's remnants still crisscross the greatly expanded campus. By the beginning of 1962, the Lunar Lab's personnel had tripled to thirty, outgrowing the corner of the PMM Building. Plans to dismantle T6 to make room for the new Science Library complicated the lack of space.

Kuiper's researchers occupied several nearby shops as the lab continued to grow, but the solution was only temporary. After two years of stopgap measures, Kuiper wrote an ambitious proposal to NASA, asking for funding for a Space Sciences Building.

These were formative years for the young research lab. Three distinct areas of research emerged: Kuiper with his lunar effort, Tom Gehrels with polarimetry, and Harold Johnson with infrared photometry.

As a young boy in Holland, Tom Gehrels had daydreamed about flight, captivated by the glamour of early aircraft. Like Kuiper, he studied astronomy at Leiden University after troubled years as a teenager embroiled in the Dutch resistance in World War II. He joined Kuiper at Yerkes Observatory in 1952 as a student assistant, already a budding philosopher unafraid to espouse unconventional ideas. He quickly became a pioneer in the field of polarimetry, which characterizes objects in space—such as dust, clouds, and planetary atmospheres—by measuring how they alter the orientation of light passing through.

Tantalized by early results on the polarization of Venus and bright stars, Gehrels envisioned a lightweight telescope that could lift above the interference of Earth's atmosphere in helium balloons. Limited science experiments had been done from balloons since the 1930s, often funded by the military. Gehrels took a test flight in a Navy gondola to prove the concept, balancing his telescope in a swinging basket in the dead silence of seven thousand feet.

But the National Science Foundation proved reluctant to fund such a dangerous endeavor, especially after a similar test flight resulted in an accidental death. Gehrels gave up the treasured idea of flying with his equipment. Instead he would develop a polarimeter that an astronomer could control from a distance. Gehrels moved to the Lunar Lab in 1961 and assembled a team to handle the complex engineering and electronics. His work rapidly broadened into three areas: ground-based polarimetry from new telescopes in the Catalina

Mountains, a balloon polarimeter called the "Polariscope," and a prototype instrument for future space probes.

Harold Johnson, a respected astronomer in the field of stellar photometry, joined the Lunar Lab in 1962. Famous for his rough-and-ready treatment of equipment, Johnson was a dynamic character in the lab's early history. He had created the international standard for measuring the magnitudes of stars with his well-established UBVRI system (ultraviolet, blue, visual, red, and infrared) while at Lowell Observatory in Flagstaff. In 1959 he moved to the University of Texas at Kuiper's invitation to observe at McDonald Observatory (which Kuiper then directed). There he decided to expand the UBVRI system further into the infrared.

Photometry is the measurement of a star's brightness—the energy it radiates in a particular waveband. Optical astronomers had mastered the art of photometry in the visible wavelengths, but the infrared is invisible to the human eye and detected only as heat.

Techniques for studying the infrared had made little progress in the last half-century. Finicky instruments had to be extremely sensitive to detect these long, low-energy wavelengths, especially because the night sky is awash with infrared radiation, a kind of continual twilight. Moreover, the Earth's atmosphere absorbs much of the infrared light emitted from space, requiring astronomers to find ever-higher mountains to obtain a clear view.

On the other hand, the infrared promised unprecedented opportunities for studying the cool bodies of the solar system, such as planets, comets, and asteroids, which mainly radiate infrared light. And because interstellar dust absorbs visible light from stars and reradiates it in the infrared, an infrared observer could peer through the obscuring haze and even gaze into the depths of the dense clouds of dust and gas that surround newly forming and dying stars.

Kuiper had made important strides in this field by working to declassify the lead sulfide detectors developed by the Germans, the first reliable instrument to measure the near-infrared. Johnson observed thousands of stars with lead sulfide detectors cooled by liquid nitrogen. He added the J, K, L, and M bands to his UBVRI system, each corresponding to a narrow "window" in Earth's atmosphere through which astronomers could make infrared observations. He needed a better detector to push his work further into the infrared.

The answer came from a physicist named Frank Low, a recent graduate from Rice University in Houston. Low had introduced himself to Johnson at the University of Texas in 1961, just before Johnson's move to Tucson. The two formed a fast friendship. At his first professional job at Texas Instruments, Low had designed an instrument called a germanium bolometer cooled by liquid helium that could measure minute changes in temperature. Johnson recognized the instrument's potential for detecting low-energy infrared photons. Together they extended the system into the N band, corresponding to ten microns in the mid-infrared.

The two briefly parted ways when Johnson moved to Tucson and Low moved to West Virginia. Johnson continued his infrared observations while Low refined the bolometer at the National Radio Astronomy Observatory with the help of a skilled instrument-builder and telescope operator named Arnold Davidson. Johnson recognized that Low and Davidson needed time on a large telescope to test the bolometer's ability to detect the infrared signals of stars.

In April 1963, Low accepted Johnson's invitation to come to Tucson for a "dress rehearsal" with Kuiper's new telescope on Mt. Bigelow in the Catalina Mountains. Johnson then arranged for the all-important test at the 82-inch telescope at McDonald Observatory in Texas.

Clouds hounded the two-week observing run that July, leaving Low and Davidson gazing at dismal skies while the liquid helium–cooled bolometer stood ready. Low recalled crossing paths with a "distinguished-looking optical astronomer" who inquired after their purpose. When they explained, the stranger responded that he couldn't understand why they were waiting for clear weather when there was nothing to see in the infrared.

At last the clouds dissipated and the humidity dropped. On a single clear night, the observers obtained measurements in the N band of the infrared spectrum of the brightness of Mars, Jupiter, Saturn, Titan, and twenty-four stars. Pleased with the data, Johnson quickly published a table of photometry in ten wavelengths, an expanded range referred to as "UBVRIJKLMN," with which he calculated the surface temperatures of different types of stars.

Johnson's infrared team included Wiesław Wisniewski, a Polish astronomer who fled his home country in 1963 during the unrest of the Cold War. Kuiper let Wisniewski know that the Lunar Lab had a position available by sending him a colored postcard, a signal they had arranged in advance. Low and Davidson joined the Lunar Lab two years later, attracted by the prospect of doing infrared astronomy among Tucson's dry deserts and high mountains.

With no space available in the PMM Building, the group set up in an empty shop on Tucson's Broadway Boulevard and continued their work. A 28-inch telescope dedicated to infrared photometry began operating in the Catalina

Mountains in July 1963. Two years later, Johnson's team added a 60-inch tele-scope to the site. Bolted to the telescopes, infrared detectors recorded data on punch-paper tape. From there the researchers transferred the data to an IBM card for processing. Kuiper bragged that the process was "so highly automated that no pencil is ever used."

LUNAR MAPPING

Kuiper himself was one of the few researchers with the tools and experience to support America's impending exploration of the Moon. Good maps of its surface were in high demand and short supply. During the first two years in Tucson, Dai Arthur and Ewen Whitaker completed editions A and B of Kui-per's *Orthographic Atlas of the Moon*, adding standard latitude-longitude lines and grids to selected sheets of the original atlas.

Kuiper wasn't entirely satisfied with the result. Because of the Moon's locked orbit, only one side faces the Earth. Craters near the edges appeared blurred and elongated in photographs, and Kuiper wanted to correct this in an adden-dum to his atlas. Begun by Arthur back at Yerkes, the task now fell to William Hartmann and full-time research assistant Harold Spradley.

"T6, being a long thin building, had the tunnel where we projected pho-tographs of the Moon onto a three-foot, white half-globe," Hartmann said. Circling round the blinding projection, he could rephotograph the globe from different directions, capturing images of the Moon's edges (limbs) as they would look from overhead. Hartmann and Spradley took hundreds of these photographs for the *Rectified Lunar Atlas*, published in 1963. Meanwhile, Arthur and Whitaker tackled the momentous task of adding and correcting the names of lunar features.

"When we projected images on that globe, we could walk around to the side and see these structures in ways that people had really never seen before," Hartmann said. That was how the Orientale Basin emerged, an immense cir-cular plain surrounded by concentric rings of mountains, overlooked by Earth-bound observers in the distortion of the Moon's limb.

Mare Orientale means "Eastern Sea." The Orientale Basin cradling the dark lava flow of the mare was indeed on the east limb of the Moon until 1961. That year, the IAU accepted Kuiper's suggestion that they rotate the traditional lunar maps so that an astronaut on the surface would see the Sun rise in the east and

FIGURE 2. William Hartmann photographs an image of the Moon projected onto a three-foot hemisphere in LPL's first premises, a Quonset hut.

COURTESY OF WILLIAM K. HARTMANN AND EWEN WHITAKER, HON. D.SC.

set in the west—a fine example of how the promise of Apollo literally turned lunar mapping upside-down.

The bull's-eye pattern, sprawled across a huge swath of the newly renamed west limb, looked hauntingly familiar. Hartmann recognized its similarity to other basins on the lunar surface, so big they hadn't been identified as massive impact craters. Projectiles had shattered the Moon's crust "like a bullet going through glass," in Hartmann's words.

With Kuiper's encouragement, Hartmann published his findings in *The Communications of the Lunar and Planetary Laboratory*. Kuiper insisted that Hartmann take credit as first author, an uncommon gesture in that rigorous academic world. Edited by British astronomer Barbara Middlehurst, the

Communications series was a throwback to an old style of astronomy, when observatories circulated self-published volumes to exchange discoveries and ideas. The slender, paper-bound volumes demonstrated another quirk of Kuiper's personality, for by that time most scientists considered prominent journals as the only valid publications.

Noble Prize–winning chemist Harold Urey, in fact, wrote Hartmann a scathing indictment of the paper, displeased both by the scientific conclusions and the journal in which it was published. "It was just a collision of two paradigms," Hartmann said, "Kuiper adopting the old idea that every observatory or research institution has its own publication." And while the *Communications* series faded away after a decade of use, Hartmann's findings on the concentric structures did not.

The *Communications* series did have the pleasant effect of liberating LPL from the fearsome "publish-or-perish" syndrome already common to many institutions. It also gave free reign to Kuiper's artistic sensibility; he liked to direct the layout and illustrations of the articles. Despite the sophisticated techniques and superior telescopes used to make the *Rectified Lunar Atlas*, Kuiper harbored a soft spot for old-fashioned hand-drawn maps of the Moon.

In 1961 Kuiper hired Alika Herring, a guitar-playing optician with an artistic bent. In addition to grinding and polishing his own telescope mirrors—widely considered among the finest available—Herring had a knack for drawing the Moon from observations and photographs. He set to work conducting seeing tests with his personal 12.5-inch telescope at Tumamoc Hill, a low peak in west Tucson. It was a modest beginning to a fine array of telescopes developed at the Lunar Lab, largely through Herring's efforts. Herring also worked on mapping the poorly understood lunar limbs by using observations from the same site.

"That was another unusual thing, a throwback to the past," Wood said. "It used to be the way people studied the Moon a hundred years ago was by making drawings. Kuiper made the first really high-quality photographic atlas of the Moon, but he was also willing to have somebody who had keen eyesight and good telescopes to make drawings as well."

THE PERFECT TELESCOPE

In 1963, a freshman named Stephen Larson found copies of the lunar atlases on the shelves of the university library. Kuiper's name on the covers caught his eye. Larson had loved the night sky ever since Comet Mrkos burned a trail above his desert home six years before. Since then, scientists who had risked studying the unpopular Moon had become commentators on America's greatest long-distance race. Unquestionably, Kuiper's lab was the place to be.

"He was Mr. Planetary Science," Larson said. "He was about the only one carrying the banner at the time."

Tom Gehrels hired Larson to help with the polarimetry program, the start of an illustrious career at the Lunar Lab. The following summer Larson plied his skills in drafting, illustration, and photography in Kuiper's lab. "It was quite interesting to be taken under his wing and have a look at what he was doing . . . which by today's standards seems very primitive, but it was cutting-edge at the time, and important in the development of science," Larson said.

Projects like these pushed the boundaries of knowledge, circumventing the traditional methods of astronomy in favor of entirely new techniques. Geology, chemistry, and atmospheric sciences became critical in ways they had never been. Yet Kuiper himself was an astronomer, and despite the advent of satellites, ground-based observing remained the only reliable way to learn about objects in the night sky.

"While our problems parallel those of geophysics, our techniques are those of stellar astronomy," Kuiper wrote in *Sky & Telescope* in 1964. "The interdisciplinary aspect of planetary astronomy requires special organizational effort."

It also required telescopes, and telescopes required mountains. "One of Kuiper's true legacies was the identification and establishment of what are considered now great observatories," Larson said. Kuiper nurtured a burning ambition to found the best observatory in the world. Flying over mountain ranges in light aircraft, he searched for lofty peaks that jutted above the scintillation of the atmosphere, where good seeing would allow crisp, steady images.

In collaboration with the University of Chile, Kuiper set up a small testing program at Cerro Tololo, a peak now bristling with telescopes. Similarly, he helped the University of Mexico develop San Pedro Mártir in Baja California as an observing site. The Lunar Lab literally paved the way for the new observatory: Arnold Evans, the observatory superintendent, bulldozed the eight-mile road to the top of the mountain, and Alika Herring carried out the seeing tests.

Kuiper's crowning success was the observatory on Mauna Kea. Despite the humid climate, the towering peaks of the Hawaiian Islands promised calm air and good seeing. In the fall of 1962, Kuiper asked Herring to take a test telescope to Haleakala, an extinct volcano on Maui rising some ten thousand feet. By this time, the Lunar Lab had built a 12-inch test telescope, but Herring insisted on fitting his own mirror into the instrument. He set up shop in an empty dome constructed by the University of Hawai'i for a planned solar observatory.

Fog spilling from the caldera on cold mornings negated Haleakala as a good site, but Kuiper soon looked toward the other peaks. In May 1964 Evans found himself once again behind a bulldozer, carving a road through ash and lava to the summit of Pu'u Poli'ahu on Mauna Kea.

With NASA funding, Kuiper built a small observatory to hold Herring's test telescope. Between June and September Herring conducted isolated observations for weeks at a time, relieved once by Hartmann, who recalled taking time off (with Kuiper's encouragement) to ramble the calderas and lava flows that so strangely mirrored the Moon.

For Herring, returning to the place where he was born meant glorious nights enthralled by the clearness of the sky. "I used to cry about that seeing," he told a friend reporting for *Sky & Telescope* years later. "During times of perfect seeing, you're held spellbound. You start to make a drawing and you forget what you were doing. I just never got enough of the Moon."

On July 20, 1964, just a month after Herring began his tests, Kuiper announced at a dedication ceremony that Mauna Kea was "probably the best site in the world" for observing the night sky. Kuiper hoped to establish a Lunar Lab telescope at the site he had worked so hard to find, but ultimately the contract went to the University of Hawai'i, which completed the 88-inch Mauna Kea telescope in 1970. Today, observatories adorn every nearby peak, with the exception of the little cinder cone where Herring first discovered the mountain's remarkable seeing.

Kuiper did not forget to search the Catalina Mountains, which rose in shades of azure and violet to the north of the university's redbrick buildings. He became intimately familiar with the summits and hollows of those mountains from the windows of a chartered aircraft. The requirements were complex. Telescopes needed clear, quiet air and little moisture, and observers wanted an accessible site that winter snows wouldn't bury.

Herring spent four months on a small knoll in the Catalinas, residing in a building recently vacated by a telephone company, to test the seeing with his portable 6-inch telescope. In January 1963, the Lunar Lab established a 21-inch reflector at the site, its dome constructed amid heavy snowfall. But Kuiper had a much grander project in mind: he wanted to build a 61-inch precision reflector dedicated to high-resolution photography of the Moon and the planets.

Later that fall, Kuiper established a flimsy, twenty-foot tower on Mt. Bigelow in the Catalina Mountains at an elevation of 8,230 feet. He supplied the tower with the 12.5-inch test telescope relocated from Tumamoc Hill. Herring spent half a year in the tower performing seeing tests. The tower was later dismantled to make room for the 61-inch telescope dome, completed in October 1965. Kuiper referred to the place as "Site I, Catalina Observatory."

NASA had agreed to fund the 61-inch telescope, though not without some protest from the astronomical community. Already, U.S. science institutions had begun to uphold an unwritten separation between astronomy and the upstart field of planetary science, the former having jurisdiction over telescopes, the latter reliant on NASA's newly developed spacecraft.

Sam Case, with his four-person staff in the Department of Physics, engineered many of the Lunar Lab's telescopes in Tucson, including the 61-inch. Robert Waland, a Scottish optician, created the telescope's main and secondary

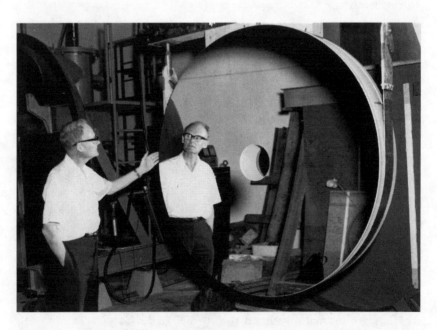

FIGURE 3. Robert Waland, chief optician, displays the 61-inch aluminized mirror for Kuiper's prized lunar observatory.

mirrors, starting his workday at four in the morning to avoid damaging fluctuations in temperature. The accuracy of the primary mirror, Kuiper claimed, was a unique achievement to rival any telescope in the world.

The creation process was not without its hitches. In 1964, the University of Arizona comptroller's office temporarily stopped construction on the telescope, and Kuiper brought in Melvin Simmons to sort out the invoices. That hurdle overcome, Simmons remained on hand to oversee the building of the dormitory near the telescope dome. The narrow redwood cabin, equipped with cots and bookshelves, provided space for half a dozen astronomers to catch up on sleep in the daytime. A steep cement pathway connected the telescope dome to the dormitory, which stood farther down the mountainside so its heat and light would not interfere with the seeing.

"We had to have steps," Simmons said, "and in the wintertime those would be covered with snow, and a tired man was liable to slide all the way down." He asked the contractors for the spare redwood left over from the dormitory and used it cover the pathway against the weather.

FIGURE 4. Gerard P. Kuiper (on right) with Robert E. Murphy
at the Catalina Station on Mt. Bigelow, January 1969

Still actively used today, the 61-inch represents an old style of astronomy, a tribute to astronomers like Kuiper who, before the age of computers, spent their nights standing on the observing platform. The graceful white dome was his showpiece, his darling dream come true. Its sturdy English yoke mount meant that the telescope could not observe above 63 degrees declination, eliminating the possibility of studying stars close to the celestial pole, but the limitation did not concern Kuiper. He intended it for lunar and planetary studies alone.

On October 8, 1965, Whitaker turned the telescope toward the Moon for the first time, with the Lunar Lab's photographer Dennis Milon on hand to take a series of trial pictures. Magnified, each crater and ridge stood out in sharp relief against the pale surface. To support the Apollo program, Kuiper's assistants spent long nights with a camera rigged to the eyepiece, patiently waiting for those exquisite moments when the atmosphere became perfectly still.

A NEW BUILDING

The telescope and dormitory complete, Simmons stayed on as the Lunar Lab's business manager, overseeing the construction of a NASA-funded Space Sciences Building. The five-story structure, completed in the autumn of 1966, made the Lunar Lab an institution in its own right. The fledging organization had finally emerged from the protective umbrella of the Institute of Atmospheric Physics, and it dropped the final part of its title to become simply the Lunar and Planetary Laboratory.

The Space Sciences Building, a long, narrow structure, contained offices, spacious darkrooms, processing labs, a library, and a vault for storing photographic plates. The elevator shaft protruded like an odd tower on the west end to preserve the option of adding a sixth story in the future. An instrument loading dock abutted the building. Inside, office doors were built extra-tall to accommodate bulky equipment—or, as Tom Gehrels later put it in his memoir *On the Glassy Sea*, to serve "as reminders to the occupants to stand tall in science."

Waland established his optical shop in a basement deep belowground to avoid damaging changes in temperature. A mezzanine level between the first and second floors housed the 80-foot globe projection tunnel for rectified lunar photography. For the analysis of planetary atmospheres, a 130-foot-long absorption tube on the third floor spanned the full length of the building, the

FIGURE 5. The completed Space Sciences Building in the fall of 1966

COURTESY OF LPL SPACE IMAGERY CENTER AND EWEN WHITAKER, SCD HC

largest in the nation. In the *Communications* series, Kuiper boasted that the Space Sciences Building even had its own small computer.

The campus had undergone other changes during the Lunar Lab's first five years. Creeping city lights forced Steward Observatory to relocate its 36-inch telescope to Kitt Peak in 1963. Kuiper objected to the plan, and in one of his typical clashes with Carpenter, insisted to President Harvill that they build an entirely new telescope. Carpenter won, however, and the 36-inch found a new home in an extra-large dome to allow for a bigger telescope in the future.

Later that year, Carpenter passed away, and Aden Meinel took his place as director. Kuiper's friend and former colleague from Yerkes, Meinel actively collaborated with the Lunar Lab, particularly in spectroscopic research. In 1964, Meinel rented temporary quarters on campus to found the Optical Sciences Center (now the College of Optical Sciences), creating a triangle of complementary research organizations at the University of Arizona.

When Meinel became director of the Optical Sciences Center in 1966, Bart J. Bok, the world's foremost authority on the Milky Way, replaced him as Steward Observatory's director and the head of the Department of Astronomy. Bok and his wife Priscilla, also a talented astronomer, came to Tucson after distinguished careers at Harvard and the Mount Stromlo Observatory in Australia. They continued to collaborate on research with a devotion to their work—and to one another—that became famous in academic circles. Bart Bok and Kuiper had been students together at Leiden University in the Netherlands in the 1920s. The two met among the musty stacks of astronomy books in the university library, where they announced their intended topics of study to each other. Both had indeed become leaders in their chosen fields.

Although they had once been friends with a sense of mutual admiration, the two Dutch astronomers often clashed as heads of competing research institutions. The rivalry continued over Bok's four years as director, the result of forceful personalities and an echo of the widening gap between astronomy and planetary sciences. Kuiper's lab, growing in eminence, would not remain dependent on telescopes, nor on the high-temperature physics and complex mathematics that shaped astronomical techniques. Spacecraft brought the Moon and planets into reach.

A few years after the completion of the Space Sciences Building, an astronomer from the University of California, San Diego, visited Tucson for a planetary conference. An impressionable graduate student at the time, John Lewis studied the chemistry of Venus with Harold Urey as his advisor. He was flattered by Kuiper's offer of a personal evening tour of the new building.

"He had his office set up sort of like a church," Lewis recalled. "There was a raised dais about a step high, where his desk was, down at the end of the room. Then he had the flag of Arizona and the flag of the United States on either side of his desk. You felt like saluting when you came into the room. I should have caught a clue from these circumstances that this was an unusual operation."

It *was* an unusual operation. Lewis wouldn't join its ranks for another decade, politely refusing Kuiper's offer of a job after Urey, unenthusiastic about his protégé's visit to Tucson, warned him to watch out for "that person." Kuiper and Urey were destined to become reluctant collaborators in America's quest for the Moon. They had few other options for colleagues. Despite growing interest in the Moon, the Lunar Lab remained one of the only places in the country actively committed to planetary research. During the lab's first five years, while Kuiper carved out a place for his work in Arizona, NASA sent rocket after rocket toward its ghostly target in the sky. The Lunar Lab, as it turned out, would be vital to its success.

Kuiper had the vision to foresee the role his little laboratory would play. As Charles Wood told the story, "I remember one rainy, rainy Tucson summer thunderstorm, he called me to his office and said, 'I need you to take this proposal we're sending to NASA to President Harvill's office at the center of campus and have it signed.' I looked out the window and said, 'But, Dr. Kuiper, there's a huge thunderstorm going on; I'll get drenched.' He said"—and here Wood made his voice deep and sonorous—"'If we want to get to the Moon, you have to walk through the rain.' And I did. And we got to the Moon, so he was absolutely right."

PART TWO

TARGET MOON

FADING PHOTOGRAPHS papered Ewen Whitaker's home office, alongside annotated Moon maps and a framed letter from President Nixon congratulating him on locating the landing site of the lunar spacecraft Surveyor 3. A lifetime of collecting and repairing old clocks had amassed into an impressive display of polished wood and delicate dials on the wall of his living room. They took a minute or two to go through their cacophony of bells and chimes.

Now in his eighties, Whitaker radiated energy and irrepressible good humor. Rumor had it he could detect ultraviolet wavelengths with one eye, allowing him to discern structures in the clouds of Venus through a telescope lens where others saw only impenetrable fog. The medical risk, his friends hastened to say, was nothing to such an indomitable astronomer.

Whitaker remembered precisely what set him down this path. He leapt up and returned to the room with a tattered copy of one volume of *The Children's Encyclopedia* in his hands. On the inside cover, an inscription in fading calligraphy read, "For your eighth birthday, June 22, 1930. Love Mommy and Daddy."

"I know it cost six pounds ten, or something like that, which was more than two weeks' salary for my father," he said, turning the book over in his hands. "They bought these for me, and I loved them, because they've got all the different subjects. There's a little bit of everything in here, about trains and people and stories in the Bible and people in Africa and maps. But in two of the volumes they go into the Earth and its neighbors, and into astronomy, and they've got little maps of the stars there, and I thought, 'Oooh.' You know, things catch your imagination when you're younger."

Whitaker set his heart on studying at Greenwich Observatory. Then came the evacuations during World War II, and with the University of London crumbling in the chaos, Whitaker never made it to college. His mother's health was failing, and he got a job at a local factory to help his father. At the Siemens Brothers Company he learned ultraviolet spectrometry to test the coating of underwater pipelines, working amid clouds of lead oxide. England needed the tough, hollow cables to secretly pump gasoline to France, and the job kept him away from the frontlines.

In 1949, Whitaker got a second chance at his childhood dream when he obtained a position at Greenwich Observatory, measuring the spectra of stars and the positions of asteroids. "I was in second heaven," he said. He joined the Lunar Section of the British Astronomical Association, a society for amateurs. In 1955, at the IAU General Assembly in Dublin, Whitaker heard Kuiper's call for improved lunar maps.

"We were of the same mind exactly," Whitaker said, "that what had gone before was nothing to lay a foundation on."

Two years later, just after Sputnik 1's launch, Whitaker flew out to Yerkes Observatory to begin printing Kuiper's lunar atlas, giving up his entire annual leave at Greenwich Observatory to make the trip. He stayed only a month, but shortly thereafter Kuiper approached him with the offer of a three-year position. Whitaker accepted, fully intending to take his wife and three young children back to England at the position's close. But by the 1960s America was headed for the Moon, and it needed Whitaker's keen eyes.

PROJECT RANGER

NASA launched the first spacecraft in the Ranger series in August 1961. Two years in the making, Rangers 1 and 2 were merely test flights, meant to pave the way for future missions. Each carried a host of instruments designed to study interplanetary space. Their failure to return scientific data dashed the country's hope of a quick rebound after the shock of Sputnik. Seven days after lifting off from Cape Canaveral, Ranger 1 tumbled out of its near-Earth orbit and burned up over the Gulf of Mexico. Ranger 2, after its launch two months later, followed suit.

In 1962 NASA launched three more Rangers, this time aimed at the Moon. Armed with seismometers, gamma ray spectrometers, and television cameras, these were "hard-landers," intended to return a burst of data in the brief minutes before impacting the lunar surface. Two missed their target entirely, and the third ceased operation just ten hours into the three-day flight.

Under fire after the successive failures, NASA reorganized Project Ranger in 1963. Administrators discarded everything except the television cameras from the scientific payload, saving the extra weight for redundant engineering features. With just one "experiment" remaining, they decided to overhaul the project's management. A team of experimenters, led by one principal investigator, would carry out scientific analysis of the photographs.

NASA assigned that role, called chief experimenter in those days, to Kuiper. The Jet Propulsion Laboratory (JPL), a research center established by the California Institute of Technology in the 1930s and adopted by NASA in 1958, would direct the Ranger missions. The experimenter team also included Eugene Shoemaker, JPL engineer Raymond Heacock, and Kuiper's archrival Harold Urey. At Kuiper's insistence, Ewen Whitaker became the team's fifth member.

The five scientists agreed on a goal for the Ranger cameras—to achieve photographs a factor of ten better than what ground-based telescopes could provide. They reserved the highest resolution for the final few frames. Shoemaker assigned Whitaker the task of finding target sites for Ranger 6 to impact. Whitaker chose Mare Tranquillitatis, the Sea of Tranquility, an area relatively free of rocks and under optimum lighting conditions for photography.

Bristling with six cameras, Ranger 6 launched on January 30, 1964. As it approached the Moon on February 2, the Goldstone Tracking Station in the Mojave Desert reported that cameras were warming up. At JPL in Pasadena, California, the team waited anxiously—"chewing our pencils," as Whitaker described it—for the next step in the sequence. The cameras never switched on. Thirteen minutes after the first images should have arrived, the tone filling the JPL auditorium abruptly disappeared. Telemetry had ceased on impact.

An inquiry revealed that a glitch during launch had short-circuited the camera system. In all other respects, the mission had performed flawlessly.

Kuiper's team had one more shot with Ranger. Plans for other missions aimed at the Moon—the Lunar Orbiters, Surveyors, and Apollo itself—were well underway, and would commence with or without the useful data that Ranger was supposed to return. Another failure would likely mean canceling the ill-starred Ranger Program entirely.

FIGURE 6. Ewen Whitaker, Gerard P. Kuiper, and Ray Heacock stand
in front of a lunar hemisphere and a model of the Ranger spacecraft.

COURTESY OF NASA/JPL-CALTECH

A new target site had to be found for Ranger 7, and the task once again fell
to Whitaker. He presented JPL engineer Don Willingham with a handwrit-
ten list of potential areas, meticulously noting rubble and other obstacles and
scrawling a "good luck" note on the back of the page. The slightly redesigned
spacecraft lifted off from Cape Canaveral on July 28, 1964.

Three days later, after a sleepless night in Pasadena, Kuiper's team gathered
in the room set aside for them at JPL, listening to reports from Goldstone
as the spacecraft drew nearer to the Moon. Whitaker had been on holiday in
England, and had taken a red-eye flight into Los Angeles for the occasion.
Kuiper, too, had just returned to Pasadena. The day before, he had made the

half-day drive to Tucson to check on the progress of his projects at home, a trip that the local newspaper lauded as typical of his energy and commitment.

Just after 6 a.m., Goldstone reported that Ranger 7 had begun to warm up its cameras. For the next thirteen minutes, video signals streamed in as the spacecraft plummeted in free-fall. At 6:25 a.m., the hum of the telemetry turned to static, and the room erupted into cheers.

"There were no pictures coming back live," Whitaker said. "We couldn't see anything, but we could hear the signal from the cameras, just a tone coming in. Then we got the first few prints—'Oh, look at this, wow, you can see the craters!' Of course the thing's photographing as it came in closer and closer, just a solid series of pictures from all these six cameras, so the view got closer and closer all the time."

Whitaker later wrote that it was the "beginning of the age of instant science." After an impatient wait for the reels of tape to arrive from the Mojave Desert, Kuiper's team had the privilege of seeing the first hurriedly produced prints late that afternoon. NASA scheduled a press conference for 9 p.m. that evening, so the team only had a few brief hours to pour over the photographs before walking into the spotlight. "This is a great day for science, and this is a great day for the United States," Kuiper said thunderously on the television screen.

Back in Tucson, the student assistants who hadn't been able to hitch a ride to Pasadena ran to the newspaper office to see the first prints. A dense bank of clouds had covered the Moon early that morning: Alika Herring, stationed at Kitt Peak to look for the impact, reported fog so thick it billowed through the observatory's doorway like smoke. Still, nobody felt disappointed. Their gamble had paid off. The study of the solar system wasn't a stepchild science any longer.

The photographs more than surpassed the scientists' goal. Alan Binder recalled, "Kuiper got up at the news conference at JPL, and in typical Kuiper fashion said, 'These pictures are not ten times better than astronomical pictures, which would be phenomenal in itself. They are not a hundred times better than astronomical pictures, which is what the engineers promised us. They are one thousand times better.' It just brought the house down."

Whitaker spent the next few weeks unrolling long 35-millimeter films and making prints from the negatives. "Oh, I tell you, two weeks of solid darkroom work," he said. "It was really an exciting time but very busy." Kuiper had agreed to undertake not only the interpretation of Ranger's results but also the monumental task of developing a loose-leaf atlas from the photos.

The experimenter team had just three months to produce a comprehensive report on their findings. Kuiper enlisted the aid of other LPL researchers to

FIGURE 7. Gerard P. Kuiper speaks at the press conference at JPL after Ranger 7's successful mission. Eugene Shoemaker and Ewen Whitaker are seated in the background while Ray Heacock looks on from the left.

COURTESY OF NASA/JPL-CALTECH

analyze the photographs. He needed the expertise of geologists—but he also hired Ralph Turner, a sculptor teaching in the University of Arizona's art department. Turner shaped accurate and artistic scale models of the lunar surface, calculating the depths of craters by scrutinizing shadows in the new images.

Ranger 7 had impacted in a shallow valley surrounded by the bright bumps of hills. At Kuiper's suggestion, the place was named Mare Cognitum, the Sea that Has Become Known.

THE LAST ATLAS

"Then things started happening very rapidly," Robert Strom said, "because we were approaching the Apollo era of sending men to the Moon."

When Strom joined LPL in 1963, in time to help analyze Ranger photographs, he added geology to the melting pot of subjects that Kuiper needed

for lunar studies. Strom remembered exactly what plunged him into the frontlines of space exploration. At a bookstore in Pakistan, where he was plying his
expertise to look for oil, Strom picked up a copy of Patrick Moore's *Guide to the
Moon*. In one edition, its cover featured a blue Earth shining above a landscape
of dark, spiky towers.

Strom thought the book was fascinating. When he returned to America, he
searched out a planetary group in Berkeley that was struggling to develop an
instrument for studying the Moon. The engineers needed a geologist to guide
them through the complexities of elemental abundances and rock formations,
and Strom's career plans took a dramatic turn. He sought out the Lunar Lab
not long after. "This was the only place at the time that studied planets, a whole
laboratory dedicated to the study of the Moon and planets," he said.

In the sixties, Strom and Whitaker acted as consultants to the Ranger, Surveyor, and Apollo programs. NASA needed geologists, not astronomers, to prepare the astronauts for what they would find on the lunar surface. Strom gave
the astronauts a crash course in geologic science, recommending features that
they should photograph from orbit.

The lab continued to expand, both in personnel and expertise. Renowned
astronomer Georges van Biesbroeck also came to LPL in 1963, beginning a
new career at the age of eighty-three. Born in Belgium, "Van B" fled his home
country for America during the German invasion of World War I, eventually
meeting Kuiper at Yerkes. His specialties—comets and double stars—required
patience and clear-sightedness that were the envy of younger astronomers. Kuiper placed a 16-inch reflector on Tumamoc Hill specifically for his old friend,
though Van B also made use of the best instruments available. He remained a
zealous and enthusiastic observer until his death in 1974.

Elizabeth ("Pat") Roemer, an eminent astronomer with a PhD from Berkeley, joined LPL in 1966. She had pursued her love of comets with determination in a time when few women studied science and even fewer obtained jobs
in a "hard science" such as astronomy. Roemer endeavored to recover returning
short-period comets and observe the arc of newly discovered ones for as long
as possible. Only after a comet had been observed twice—once on its discovery,
and once on its return—could an astronomer calculate its period (the time it
takes to complete one orbit) and thus predict its next passage into Earth's view.

Quiet, unassuming, and immensely intimidating to students, Roemer spent
countless nights capturing images of comets. She tracked their motion on photographic plates where they appeared as small round spots against a backdrop

of star trails. With those images she could evaluate a comet's origin and chart its path through the solar system.

Roemer also estimated the dimensions of comet nuclei, generally only a few kilometers across—a tricky task because the bright core is shrouded by the glowing coma, a fuzzy haze of gas and dust. Her catalog of nuclear magnitudes remained the most systematic effort in this area of research until the advent of charge-coupled device (CCD) cameras in the eighties.

"Only a very few people had the resources to make regular observations of faint or distant comets," Roemer said. "Thus the observations I made were often the first, as well as the last, of returning periodic comets at successive perihelion passages. And I was often the last to observe newly discovered comets as they faded with increasing distance from the Sun."

In the midst of this expansion of stellar research, Kuiper made sure the Catalina observatories remained available for lunar studies. In the photographs published in the *Rectified Lunar Atlas*, previously unseen features had appeared around the edges of the Moon. Dai Arthur, assisted by Ruth Horvath, Charles Wood, and Alice Agnieray, spent five years updating the IAU's 1935 catalog of lunar formations. The new map was printed in LPL's *Communications* in 1966 and approved by the IAU the following year.

Kuiper had nearly fulfilled his ambitious plans for the lunar atlas. He still hoped to publish a final version containing the highest-resolution photographs ever obtained from the ground—the images from his beloved 61-inch telescope in the Catalina Mountains. Whitaker, Strom, John Fountain, and Steve Larson selected and prepared 225 prints for the *Consolidated Lunar Atlas*, completed in 1967.

Kuiper distributed copies to libraries and universities, but the atlas was never released for general sale. Apollo had overshadowed Kuiper's dream. Telescopic observations of the Moon could not compete with the detailed images returned by spacecraft. Whitaker wrote that it was "the finest lunar atlas ever produced from ground-based photography, and probably the last."

SEAS OF DUST

NASA launched two more Rangers in February and March of 1965. The success of Ranger 7 took the task of choosing landing sites out of Whitaker's hands. Apollo officials claimed the right to select targets that would help smooth the way for astronauts. Ranger 8 successfully impacted in Mare Tranquillitatis, and the press received photographs the same afternoon.

Emboldened by success, the experimenters agreed to command Ranger 9 to perform a terminal maneuver that would improve the resolution of the photographs. The radio antennae had been modified so that the images could be picked up by commercial broadcasts. The spacecraft's plummet into Crater Alphonsus was televised live, offering front-row seats to the eighteen-minute event in living rooms across America.

Like Sputnik's launch a decade earlier, the moment would inspire kids around the nation to study space. "The thrilling thing was not just seeing the Moon coming at you—because they had the first picture, the next picture closer up, the next picture close up—but below they said, 'Live from the Moon,'" recalled Guy Consolmagno, a thirteen-year-old boy at the time growing up under the dark skies of Michigan, who would later end up at the Lunar Lab.

Lunar Orbiter 1, the first of five, transmitted its photos in August 1966. Methodically mapping the Moon, including nearly all of the unseen far side, the Lunar Orbiters searched for landing sites for the all-important Apollo missions. For scientists who studied the Moon, the three years following Ranger 7's success were astonishing. "Big boxes of these images would keep coming in every few days," Whitaker said. "We were like kids in a candy shop, seeing all these new formations at high resolution."

But the early decision to include only photographic equipment took its toll. Scientists remained uncertain about the nature of the lunar surface, a puzzle that had to be solved before Apollo could proceed with any confidence. The hard-landing Rangers had not attempted to settle gently on the surface, as a manned lunar module would have to do. Unable to wait for the long-delayed Ranger data, engineers had designed the Apollo landing gear with solid ground in mind, at that point little more than an optimistic guess.

"We were trying to extract as much information as possible just from imaging," Stephen Larson said, "and there was a lot of contention at the time about whether or not the surface was even strong enough to sustain the landing of a spacecraft. Some people predicted it was a very loose, powdery thing that would just swallow [the spacecraft] up when we tried to land."

The idea, unlikely as it seemed, had the potential to put a stop to NASA's plans for manned exploration. Its main champion was Thomas Gold, a respected astronomer from Cornell University. Clearly the lunar surface was dusty, but should the dust be measured in centimeters, meters, or kilometers? Gold hypothesized that the dark areas of the Moon were in fact "seas" of fluffy dust that would engulf a soft-landing spacecraft without leaving a trace.

The question of the Moon's composition and origin had created a decade-long rift between Urey and Kuiper ever since their first snappish dialogue via scientific papers in 1954–55. Urey theorized that the Moon was an accumulation of rocky debris, a "cold Moon" without any active volcanism in its history. He campaigned for a sample return mission, convinced that images alone could not provide the answer.

Kuiper, however, had chartered flights over volcanic regions of Mexico, Hawaii, and the southwestern United States as part of his efforts to analyze the Ranger photographs. The trips reinforced his impression of the similarities between the Moon's maria and dark, glassy fields of lava. He maintained his position that the young Moon had melted, leaving igneous rock behind.

Apollo samples would settle the issue, but NASA first needed proof that the lunar surface had the strength to bear a soft-landing spacecraft. On March 24, 1965, Kuiper arranged for Herring and Wood to observe the impact of Ranger 9 from Kitt Peak's 84-inch. It was almost unheard of to use such a large telescope to view an object as bright as the Moon, but if Gold's theory was correct, an observer should see a large dust cloud puff up at the moment of impact.

"Almost nobody, then or now, looks at the Moon through a large telescope with their eyeballs," Wood said. "I remember the stability of the atmosphere for seconds would be very good, and we could see tiny, tiny craters on the Moon that no one had ever seen before."

In the stillness of the desert morning, they waited for word on the radio that Ranger 9 had hit. The result—absolutely nothing—was one of the most encouraging signs that "Gold dust" could be disproved. Later Kuiper calculated the bearing strength of the lunar surface based on rocks ejected from an impact crater photographed by Ranger 9. His conclusion allayed NASA's fears that future spacecraft would be swallowed into the unfathomable depths.

Americans would later marvel at the footprints of Neil Armstrong and Buzz Aldrin in the fine dust of the Moon. Those photographs became a stunning representation of humanity's resourcefulness. But for LPL scientists, that footprint was much more personal. It was the long-awaited result of their efforts to prove that the Moon would bear an astronaut's weight.

INVESTING IN INFRARED ASTRONOMY

In this age before frequent spaceflight, many scientists at LPL remained dependent on ground-based telescopes for their research. The Ranger missions, immensely exciting to the press and the public, remained only a small component of LPL's activities. To support the work of astronomers like Roemer and Van B, as well as the new program of infrared astronomy, Kuiper expanded and rearranged his telescope facilities in the Catalina Mountains.

At Site I on Mt. Bigelow, Kuiper's 61-inch telescope peered beyond the close orbit of the Moon to see red-banded Jupiter and capture white clouds across the ochre surface of Mars. Kuiper eventually returned the nearby 21-inch telescope to its original location on Tumamoc Hill, conveniently close to campus and already hosting Van B's 16-inch reflector. He replaced the 21-inch with a graceful Schmidt telescope, ideal for planetary photography. Roemer and Van B spent long hours at these telescopes, tracking the paths of comets across the sky.

Harold Johnson and Frank Low, breaking new ground in the infrared region of the spectrum, also needed high platforms to make observations. They adopted "Site II" in the Catalina Mountains, half a mile away from Site I and two hundred feet higher. The first Site II telescope was Johnson's 28-inch, dedicated to infrared photometry.

Many astronomers still believed that there was nothing to see in these long, low-energy wavelengths. In those days, infrared observers had to bolt their equipment onto borrowed optical telescopes, a practice tolerated by other astronomers because infrared observations could be done during the otherwise useless bright time. These temporary fixes, however, made it difficult to remove the "sky noise" created by the warm and turbulent air inside telescope domes.

Johnson envisioned a suite of telescopes specially designed for working in the infrared. The 28-inch could "wobble" rapidly between a star and the sky as the detector recorded measurements, allowing observers to subtract the large infrared signal given off by the sky and see only the targeted object. Johnson housed the telescope in a secondhand building with a roll-off roof that gave observers a wide field of view.

Robert Waland constructed three 60-inch aluminum mirrors in LPL's optical shop, following the design of the 28-inch—inexpensive, with less precision than an optical telescope, but sufficiently capable of feeding the infrared image of a star into the detector's large aperture. Nine inches thick in the center, these "no frills" telescopes could be manually unclamped to rapidly scan the sky. Johnson called them "a compromise between cost and precision."

The prototype 60-inch went to the new observatory at San Pedro Mártir, and the others were installed at Site II in the late sixties, one for Johnson's use and one funded jointly by the University of Minnesota and UC San Diego. Johnson also made use of a 40-inch telescope established on nearby Soldier Peak in 1969.

The program prospered in the dry desert air. Johnson and Low published the first authoritative measurements of stars in the near-infrared and even captured the hint of a quasar (a brilliant jet stream powered by the black hole at a galaxy's core)—the first object outside of the Milky Way to be detected in the infrared. Graduate student Doug Kleinmann stumbled on a nebula shrouded with dust in the visible wavelengths, but so bright in the infrared that it saturated the instrument. Kleinmann reportedly told Low about the discovery ashen-faced, certain he had broken the detector. There was plenty to see in the infrared wavelengths, after all.

Kuiper's willingness to make room for people like Johnson and Low created a space for unconventional research within the lunar-focused lab. Johnson joked that he was in the "stellar division" of LPL. However, other departments soon lured him away with the promise of pursuing his interests free from Kuiper's autocratic oversight. In 1969, he transferred to the Optical Sciences Center, and then to Steward Observatory. Simultaneously, he began working part-time at the National University of Mexico to help develop the new observatory at San Pedro Mártir. He encouraged LPL to transfer the first 60-inch telescope to the site, and he eventually moved to Mexico to continue his research there.

Johnson's legacy impressed George Rieke, a Harvard physicist who joined LPL in 1970. In his relatively brief time at LPL, Johnson had created a leading infrared photometry program and took the first tentative steps into the study of infrared spectroscopy. "It's incredible that in four years he went from a clean sheet of paper to a mature area of astronomy," Rieke said.

Rieke recognized that Kuiper actively sought scientists who did what he called "slightly offbeat but very technically advanced kinds of astronomy." When Rieke joined the infrared group as a postdoctoral associate, transferring his research interests from gamma ray astronomy, he was also impressed by Low's forceful personality and technical expertise. "He taught me what a genius is, because he had intuitive leaps to really good technical solutions," Rieke said.

On the east end of the Space Sciences Building's first floor, the group continued building increasingly sensitive detectors. The researchers weren't tied to any one area of astronomy—planets and stars made equally valid targets. Low encouraged his colleagues to observe rocky objects and generated a thermal map of the lunar surface with an infrared instrument attached to the Apollo 17 orbiter. Rieke, meanwhile, focused his research on galaxies. The instruments revealed a particular type of luminous galaxy that formed a core of hot, young stars within a veil of cool dust. Although these "starbursts" turned out to be quite common, astronomers could only detect them by observing in the infrared wavelengths.

"It was like an open playing field," Rieke said. "You could build something, and then just go observe anything you wanted to, and there was a pretty good chance that something interesting would happen."

Modifications to the Site II telescopes continued. One clear evening in Tucson, Low set up a small telescope in LPL's parking lot and experimented with tilting a secondary mirror manually, moving the star on and off the detector. This turned out to be an excellent way to remove the infrared background and greatly reduced the troublesome sky noise from the telescope dome. Secondary mirrors were soon added to the 28-inch and 60-inch infrared telescopes. The telescopes also got a higher platform when the Air Force Command vacated a radar base on Mt. Lemmon. Kuiper acquired a permit to use the land and transferred all the infrared telescopes to the taller summit in 1972, returning Site II to the U.S. Forest Service.

The superbly outfitted 61-inch telescope on Mt. Bigelow promised even better results. Its design allowed modifications that would eliminate the thermal emissions from the telescope, allowing the detector to see only the sky. Initially, Kuiper felt reluctant to allow any tinkering on his prized lunar observatory. When he relented, Rieke and Low retrofitted the 61-inch with an alternative mirror to combat the noise emanating from the telescope's warmth. Kuiper's gem, as Low called it, became the first ground-based telescope fully optimized for infrared observing.

By this time, the California Institute of Technology (Caltech), University of Minnesota, and UC San Diego had developed their own pioneering infrared groups. This made it possible for LPL to confirm discoveries with other observatories, although the exquisitely prepared 61-inch telescope could sometimes detect things that Caltech's 200-inch couldn't find. In the spirit of cooperation, Low established a private company, Infrared Laboratories, to make his detectors available to other observers.

"You wouldn't tell the deep secrets of how you did something, but you would show enough that people could benefit from what you'd done," Rieke said. "In some ways the rivalry was fiercer than at present, but in other ways it was much more gentlemanly—the way you imagine science should be done."

ABOVE THE ATMOSPHERE

Even on the tallest mountains, water vapor in the atmosphere often swamped the faint signatures left by planets. Scientists fashioned innovative solutions to get their telescopes above the turbulent air. Tom Gehrels enlisted the help of the National Center for Atmospheric Research (NCAR) for his Polariscope program. NCAR had recently established a balloon launch center in Palestine, Texas. In the early sixties they released test balloons from an alternative launch site beneath the shelter of Glen Canyon Dam, its lake bed not yet filled.

"It was like being down in a box," reflected Dale Cruikshank, who visited the site with Kuiper to send aloft an infrared instrument. "It was an ideal place to set up a balloon, which would inflate and rise while it was still tethered to the ground. You could get all your equipment working and then cut the cord, and it would float up out of the canyon in front of the dam."

Gehrels flew the Polariscope on its maiden flight in 1966 under NCAR's watchful eye. Balanced precariously in the aluminum latticework of its gondola, the 28-inch radio-controlled telescope drifted beneath a gauzy helium balloon at an altitude of 120,000 feet. The Polariscope flew four times in all, making polarization measurements of stars at wavelengths in the deep ultraviolet. Predicting its path was often a challenge, even with the light surface winds at Page, Arizona, or Palestine, Texas. Despite a near disaster when the gondola released from its balloon precariously close to the Grand Canyon, Gehrels recovered the precious cargo after each launch.

Gehrels continued to make polarimetry measurements from the Catalina observatories, but he had even more ambitious heights in mind. He intended to design a polarimeter that would fly on future spacecraft missions. Polish astronomer Krzysztof Serkowski joined his project in 1970, driven by the dream of discovering planets around other stars. The group also included a young Jesuit priest named George Coyne and former engineer Martin Tomasko. They discovered polarized light emitting from certain stars and galaxies, and the program rapidly broadened from solar-system objects into the far reaches of space.

FIGURE 8. Kuiper's balloon launch from Glen Canyon, November 1963

COURTESY OF D. C. CRUIKSHANK

Infrared observers also had to contend with water vapor, even on the sites high in the Catalina Mountains. Parallel to the ground-based program, Frank Low hoped to develop high-flying aircraft as a solution. Carl Gillespie, a retired Air Force pilot, joined the group in 1966. Gillespie obtained an old Navy bomber, and the team fitted a 2-inch telescope into its porthole, targeting the Sun. At 40,000 feet, Gillespie flew in his shirtsleeves.

Low flew his helium-cooled bolometer on fourteen flights, making a precise measurement of the Sun's surface temperature in the far-infrared. When the plane and its two pilots were lost on a transcontinental flight, he went in search of a replacement. At the NASA Ames Research Center in California, he located an old Learjet and partnered with an infrared group at Rice University to fit a 12-inch telescope into its emergency door.

A pilot named Glen Stinnett Jr. volunteered to take Gillespie on the first test flights to prove that the modifications hadn't interfered with the safety of the aircraft. That accomplished, the first observing flights took place in October 1968. The two pilots and the telescope operator wore breathing apparatus so that they could reduce the cabin pressure to allow the instrument to swivel freely in its bearing. The high altitude eliminated much of the Earth's atmosphere, and Low invented the chopping secondary mirror during this time to subtract the sky's brightness. Low, Gillespie, and two Rice University graduate students named George Aumann and Al Harper made roughly eighty-five flights on the Learjet over the next three years. Scientific advances included the first measurements of the internal heat of Saturn and Jupiter.

Meanwhile, Kuiper was engaged in a similar effort to locate a plane for taking the spectra of planets using Johnson's new infrared spectrometer. Johnson had introduced Fourier transform spectroscopy to the lab, which involved using an arrangement of mirrors to collect a signal simultaneously at many wavelengths and a computer to process the raw data into a spectrum.

When Johnson left LPL, Kuiper hired physicist Harold Larson to continue the airborne spectroscopy program. As a postdoc in France, Larson had developed similar spectroscopy techniques for ground-based telescopes. His contact with Kuiper prior to accepting the job was limited to two letters. "I like to say I was a mail-order professor," Larson said. "I hadn't made any visits; I had no idea what I was getting into. . . . I didn't think I could name the nine planets. Kuiper said, 'It doesn't matter. You learn planetary science by doing it.'"

What he was getting into was a twenty-five-year-long involvement in airborne spectroscopy. NASA had just purchased a Convair 990 (dubbed "Galileo") to use as an observing platform, open to scientists around the world. Larson's group included an expert in planetary atmospheres, Uwe Fink; a colorful Austrian engineer named Ferdinand DeWeiss, who designed the telescope's support system; and Dale Cruikshank, who tested the onboard spectrometer.

Fink had earned his PhD from Pennsylvania State University, where he had made one of the first measurements of the amount of methane present in Earth's atmosphere. He was interviewing for a job in environmental work when Kuiper invited him to come to the Lunar Lab, taking Fink up in the CV-990 as an enticing lure.

"Kuiper told me, 'spectroscopy is the queen of astronomy,'" Fink recalled. Although focused on his lunar effort, Kuiper's first interest and prodigious skill remained in analyzing the spectra of planetary atmospheres. "At that time this was a wide-open field, you know, because we did not know much about the planets at all," Fink said. "We didn't know the surface pressure of Venus; we didn't know how much atmosphere there was on Mars; we didn't know most of the constituents in the atmosphere on Jupiter. This was really the beginning of planetary science."

Larson's group took a two-pronged approach to the study of planetary atmospheres. First, they would collect measurements of light from the high platform of the CV-990. Next, to see which gases matched those spectra, they experimented with the absorption tube (called a White cell) that spanned the length of the Kuiper Space Sciences Building. The stainless steel tube, 130 feet long and 2 feet wide, could hold up to ten atmospheres of pressure—though nobody ever tested that limit. A beam of sunlight, bounced back and forth between mirrors to simulate the long length of a planet's atmosphere, passed through the mixture of gases pumped into the tube. By comparing spectra taken on flights with results gained in the lab, the researchers could infer what molecules made up a planet's atmosphere.

"It wasn't dangerous if you knew what you were doing," Fink said. "We did fill the tube with several atmospheres of hydrogen and methane, clearly very explosive mixtures, and every time we pumped it out we vented it to the outside. We didn't have to ask permission."

Unlike the Learjet, which had an open port, the CV-990 had thick quartz windows that limited the transmission of infrared light. Despite this drawback, the group made important discoveries, not the least of which was how to operate a telescope in what Larson called "a very hostile environment." In May 1967, Kuiper announced that their data disproved the common assumption that water vapor filled Venus's atmosphere.

"We were plucking all the easy things," Larson said. "But back then, none of us really felt comfortable. We were always pushing the limit of something and never knew what was going to happen, and always surprised and amazed that we were achieving results that got national attention. It was a privileged time to be working in science."

In April 1973, while returning from an unrelated project, the CV-990 collided with another plane near Moffett Field, killing 16 people. By then, plans for a bigger airborne observatory were underway. NASA provided a larger plane, a Lockheed C-141 Starlifter. Although built with the experiences of the Learjet and CV-990 in mind, this new plane outshone its humble beginnings. It carried a 36-inch telescope that peered from a retrofitted portal, and it could carry twenty scientists. Flights lasted far longer than the tiny Learjet could manage. Best of all, a bulkhead separated the crew cabin from the open porthole, making oxygen masks unnecessary.

The C-141 flew an average of seventy flights a year over the next two decades, until NASA grounded it to make room for a more sophisticated high-altitude observatory. Over its lifespan, it contributed prolifically to knowledge of solar system objects. Larson wryly recalled his first uncomfortable flight. The plane lacked insulation, so the scientists hunched in the freezing shell, listening to the roar of the engines. It had no windows to break the hours of monotony. At last his turn for observing arrived, and the boredom gave way to frenzied activity.

"Observing isn't pleasant," Larson said. "It's hard work, it's tiring, and 80 to 90 percent of what you do is never useful for anything. But then you get a discovery. Then you forget all the bad moments, because suddenly something is important. We had enough discoveries to make the whole effort worthwhile."

THE SEARCH FOR SURVEYOR

Surveyor 1, the first of seven soft-landers sent to the Moon, was a test of engineering. Later Surveyors would include soil-analysis equipment and magnetic footpads, but Surveyor 1 carried only a television camera. Its successful landing on June 2, 1966, unequivocally proved that the lunar surface would not swallow unsuspecting astronauts into chasms of dust.

Exactly where the landing took place, however, wasn't clear. Surveyor 1 touched down somewhere in the plain of Flamsteed P, a broad crater enclosed by lava flows. The spacecraft cobbled together a panorama that showed two bright peaks on the rim of the crater. JPL engineers assumed these peaks had caused two anomalies in the landing radar and used them to estimate the spacecraft's position. They published their results in *Science* later that June.

Ewen Whitaker, now a member of Gene Shoemaker's imaging team, puzzled over their conclusion. Comparing the hills in Surveyor's hodgepodge mosaic with the best ground-based photos, he triangulated Surveyor's location. The estimate he arrived at didn't match JPL's results. Something other than the peaks must have reflected the landing radar.

"Lo and behold, looking around, there were two very bright tiny craters," Whitaker said. "I thought, uh-oh, I bet those were the two things that caused the blips in the radar." He explained his method in the September issue of *Science*. Photographs from Lunar Orbiter 1 later showed Surveyor 1 within a hundred yards of Whitaker's prediction.

"That got me that little job," Whitaker chuckled. His detective work ultimately located the landing sites of four Surveyors, gleaming like proverbial needles in the immense cratered plains. The years 1966–68 saw a remarkable record of success in American missions to the Moon, salve to the sting of the Ranger Program's initial failures. Five Lunar Orbiters scoured the Moon for Apollo landing sites and mapped the unseen far side, while five successful Surveyors (out of seven attempts) photographed their footpads pressed delicately into dust.

The honor of determining the Moon's composition fell to Surveyors 5, 6, and 7, each equipped with an instrument that could measure the abundance of elements in the Moon's regolith (a term newly applied by Shoemaker to the fragmented material covering the lunar surface). The argument between Kuiper and Urey had remained hostile throughout much of the sixties, though by the time Surveyor 5 returned the first scientific data in favor of lava-formed basalts, they reportedly could hold amiable conversations. Urey held out hope for his cold Moon theory until Apollo samples finally settled the question: the lunar maria was composed of igneous rock formed in the few billion years of active volcanism following the Moon's molten birth. To Urey's credit, he recanted his earlier theory and adjusted his conclusions.

Science, however, took a backseat to the practical matter of determining landing sites for the Apollo missions. That lent some urgency to Whitaker's task of locating lost Surveyors. He earned accolades from President Nixon for finding Surveyor 3, which alighted somewhere within the dark rubble of Oceanus Procellarum, the Ocean of Storms, in April 1967. Angled edges in the grainy photographs that Surveyor 3 sent back suggested that it had landed inside a crater.

JPL presented Whitaker with three long strips of Lunar Orbiter photographs, taken before Surveyor 3 touched down. He scrutinized the pictures with a magnifying glass, trying to identify a defining feature that matched Surveyor 3's view. Twenty-two hours later, he pinpointed the small, unnamed crater in which the spacecraft had settled.

Today's space missions launch with more glamour and glitz, their paths to Mars, Saturn, Pluto, and even beyond the tenuous reach of the Sun's influence tracked with scrupulous precision. Yet the field of planetary science grew out of work done with pencils, darkroom chemicals, and the sharp eyesight of a select few whose passion mattered more than their training or nationality. The Rangers, Surveyors, and Lunar Orbiters blazed the route to the Moon. At the end of the decade, President Kennedy's vision for space exploration çame to pass. The first astronauts set foot on the lunar surface. The role of scientists began to fade as engineers and pilots took the helm.

Whitaker, who had begun his career in England as an amateur astronomer, had a proud role to play in America's first foray into space. When Apollo 12's lunar module settled in Oceanus Procellarum on November 19, 1969, in the slow-moving, shadowy border between the long lunar night and day, Surveyor 3's little crater could be glimpsed within easy walking distance. After their first brief tour of the lunar surface, astronauts Pete Conrad and Alan Bean spent a "night" of resting and planning before walking to the rim and peering down in hopes of seeing some sign that humankind had placed its metal footprint there before their arrival.

"I met those astronauts later," Whitaker recalled. "'Oh, you're the guy who found the Surveyor,' they said." On their tour of the lunar surface, they had seen sunlight glinting off the dust-coated lander, exactly where Whitaker predicted it would be.

MEN ON THE MOON

Alan Binder grew up in a tiny farming community not far from Yerkes Observatory. In one of his earliest memories, he dozed in the backseat of his parents' car on the way home from his grandparents' farm. Looking out the window as fields blurred past, the multitude of lights in the sky dazzled the six-year-old. He turned to his brother—older and wiser by four years—and asked, "What are those things?"

"I have no idea what he said," Binder laughed. "He probably said some nonsense. But ever since then, I was interested in astronomy and the Moon and the planets."

Years later, Binder saw Kuiper interviewed on television, standing on the balcony of McDonald Observatory in Texas. Impudently, he introduced himself in a letter, a contact that grew into a summer position at Yerkes. He followed Kuiper to Tucson to pursue a PhD in geology while working at LPL.

His assistantship to the illustrious Dutch astronomer was not a smooth ride. Kuiper's indifference to the students often frustrated and disgusted him.

Nevertheless, Binder's overwhelming emotion during the heydays of Apollo was pride. "I wanted to explore, I wanted to get my feet on the ground," he said. He dreamed of studying the solar system the way natural scientists once studied Earth's landscapes, seeing each crater for the first time on a level view. When the lunar module of Apollo 11 touched down at Tranquility Base, he watched the ghostly images on the television at the edge of his seat.

"Your heart was skipping," he recalled. "God, we're down! Get the rocks, get back in and make sure you get back. It was so new and it seemed so dangerous that your heart was in your mouth because you wanted it to succeed."

NASA launched seven crewed missions to the Moon between 1969 and 1972. Only one did not reach its destination—Apollo 13 returned its crew safely to Earth after a malfunction forced NASA to cancel the lunar landing. Along with its astronauts, Apollo 14 carried hundreds of seeds that later sprouted into "Moon Trees" distributed around the country. One, a knobby sycamore, grows outside the door of the Space Sciences Building. All told, the astronauts brought back 382 kilograms of lunar rocks and regolith, which NASA distributed to labs around the world. Analyzing the samples would keep researchers at LPL busy for years.

Kuiper's involvement with the lunar program had put Tucson on the map. The night sky took on new meaning. "You could look up, and people were up there," Binder said. "Because I had my telescope, I would look at the landing site and see the mountains and see the craters and I knew there were people there. I could look in the window at the television and see those mountains. That was an amazing connection to me. . . . I have all these fantastic memories of Apollo and the men on the Moon, and I envied them so much because I wanted to go."

Many students shared that desire. It seemed that in no time spacewalks and moonbases would become commonplace. Hartmann remembered walking across the campus and looking up to see the daytime Moon, a crescent sharply defined against a blaze of blue. "I'd be thinking to myself, gee, it's only going to go around maybe sixty times before we actually try to land on it," Hartmann said. "It was kind of a personal relationship with the Moon."

Other scientists at the Lunar Lab shared that sense of closeness and possession. In the early days, Kuiper speculated on volunteering for a one-way trip to the Moon, providing he could communicate with Earth for at least an hour.

"We all wanted to study the Moon and the planets," Binder said. "The whole world was listening." He believed he would complete his PhD thesis on the Moon, studying its composition and history not through the impersonal distance of a telescope lens, but by touching and walking on its surface.

That dream never came true. Pilots and engineers filled the crew manifests for NASA missions, often leaving scientific research on the ground. Increasingly, scientists grappled with the concept of designing instruments that worked without human aid. The old style of observing, where an astronomer stood in the frigid telescope dome and manually moved the gears, no longer applied. Automated spacecraft were their ticket into the space program.

In December 1972 the last lunar module touched down in Mare Serenitatis, the Sea of Serenity, with its precious human cargo, including Harrison Schmitt, one of Shoemaker's Flagstaff geologists and the only scientist to ever walk on the Moon. In the decades since, the Moon has once again fallen into disfavor. Even before the last lunar landing, NASA began preparations to reach more distant destinations with the Mariner and Pioneer series.

GERARD PETER KUIPER

"The Lunar Lab is not what it was then," Binder said. "Then it was Kuiper. That was it. When one says to me, 'Lunar Lab,' my mind just goes back to old Gerard."

Colleagues remember Gerard Peter Kuiper—or "GPK" as they called him, though to students he was always "Dr. Kuiper"—as quaint, old-fashioned, and sharp-minded. A forceful and penetrating personality, he wore a jacket and tie to work every day, insisting on a formality and decorum that had faded away from other scientific institutions. His front pocket always contained a short stub of pencil with which to illustrate his ideas on whatever scrap paper he could find.

Born in the Netherlands as Gerrit Pieter Kuiper, he carried with him the authoritative style of European universities. He was undisputedly the head of the Lunar Lab, its benevolent dictator, the calm center of all its activities. As a driven and self-possessed student at Leiden University, Kuiper had already chosen the solar system as the focus of his research. His incoming class in 1924 consisted of only one other student, Bart J. Bok, who was destined to become Kuiper's friend and rival in Tucson, situated just across the street at Steward Observatory.

When the two met beside the astronomy shelves in the library, Bok announced his intention to focus his research on the Milky Way. As Bok recalled, Kuiper immediately replied, "That is not an uninteresting field. But I expect to study a more fundamental area, the problem of three bodies and related questions about the nature and origin of the solar system."

Despite his early intentions, Kuiper found it nearly impossible to start his career by studying the disregarded solar system. Instead he turned to binary stars, hoping to find an analog for planetary formation. At the University of California's Lick Observatory, Harvard, and finally Yerkes, Kuiper delved into star clusters, white dwarfs, and the luminosity of stars. World War II interrupted his research, and he developed radar countermeasures in wartime service. He obtained leave to visit McDonald Observatory during this time, where he took his first steps into planetary astronomy, detecting methane on Titan and proving that a satellite could have an atmosphere. He discovered Uranus's moon Miranda in 1948 and Neptune's moon Nereid the following year.

In the late forties, Kuiper made the shocking prediction that the universe contained many suns, each with a family of circling planets spun out of swirling clouds of primordial gas. At the time, astronomers favored the theory that the solar system had formed from rare cosmic collisions, making the Earth exceptional, not commonplace. Kuiper, on the other hand, thought one out every hundred stars might harbor a planet, a calculation that withstood the test of time.

Kuiper's desire to study the origin of the solar system led him to search for clues in the nearby Moon. At the Lunar Lab, he could pursue his first interest with a characteristic single-mindedness. He relocated to Tucson with his wife, an American named Sarah Parker Fuller, and two children, Paul Hayes and Sylvia Lucy Ann. Colleagues remembered Kuiper as a devoted family man. He delighted in tending the jungle-like garden at their house on Sawtelle Avenue.

Kuiper approached science with an artist's eye. He was not blind to the importance of his work. The eastern side door of the Space Sciences Building remains locked to this day, an unremembered tribute to Kuiper's command to keep the public from trafficking through his halls. "He was a very, very demanding individual," Dale Cruikshank said. "He worked extremely hard himself, and he demanded the same dedication, devotion, seriousness from everybody around him."

Research assistants, struggling to keep up with the indomitable astronomer, spent long nights in the telescope dome. Catnapping on the cold floor of the observatory, Kuiper could wake twenty minutes later, refreshed and ready to go back to work. A versatile innovator, he printed his own photographs in the darkroom and once used a bicycle chain to construct a scientific apparatus to study Pluto.

To students, he seemed imposing in person, intensely devoted to his work—and to his beloved telescopes—almost to obsession. A research assistant, Lyn Doose, remembered the longest conversation he and Kuiper exchanged: "I was observing up at the 61-inch, and he was very worried that I was going to back my car up over his flowers that he planted up there."

Kuiper circulated formally composed memos complaining of any irregularity that threatened the 61-inch, from unlocked storage sheds to tourists siphoning gas from the company vehicle. George Rieke remembered one such reprimand: "I once left the lights on in the tunnel going down to the dormitory at the 61-inch and I got written up in a memo that got distributed to the whole lab." Kuiper eventually installed a lock that only a privileged few could open.

Isolated in his position as the autocratic head of LPL, Kuiper often craved an audience for his thoughts. With old-fashioned courtliness, he dictated out loud to his longtime secretary Ida Edwards or called on passing students to provide a respectful ear. Chuck Wood recalled waking early on Saturday mornings to a summons from Kuiper's assistant.

"I'd go over there and [Kuiper] would ask me a few questions about something, and then he'd start storytelling," Wood said. "He told me about after World War II when he was trying to find von Braun, the German rocket designer. He had part of a German rocket motor that von Braun had built in his office, which he showed me. It was really amazing to have him need an audience, and I was the audience."

Kuiper had immense faith in the U.S. space program, expressing in a letter to University President Harvill his admiration for its "spirit of discovery" and "complete openness of communication."

To balance that, he felt deep concern over the social and political unrest that marked the sixties, to the extent of calling Wood, whom he considered LPL's resident hippie, into his office one day to ask if students might attack the government-funded Space Sciences Building.

Kuiper had little patience with naysayers of space exploration. He complained to Harvill that they blithely took advantage of technological comforts without any appreciation for the rigorous intellectual disciplines that made such technology possible. "I see no alternative but to teach more science, philosophy, and particularly the scientific methods of problem-solving to our students in law, government, and humanities," he wrote in 1972, anticipating a crisis in science education that continues to spark similar pleas today.

His concern for the state of education did not make him a personable teacher, however. He had a talent for inspiring the public, but students who searched him out found him a difficult boss. Dignified, polished, and arrogant, Kuiper was prone to sparking ruptures between his colleagues. As a measure of his personality, he sometimes wrote letters to prominent authors to express admiration of their work, in which he couldn't resist pointing out errors in mathematics or spelling. Cruikshank recalled that Kuiper could be abrupt and dismissive at times, but also said that he learned a sense of "elegance and beauty" from Kuiper as they watched the spectra of stars emerge, line by line, from an instrument strapped to the telescope.

Many who knew him recalled a warm and generous person beneath the redoubtable scientist. He hosted going-away parties for students at his home and made allowances for slacking at their work if they'd newly fallen in love— though missed classes and sleepless nights were often required to keep up with the demanding assistantships. "Even though he was intimidating to us," Hartmann said, "he had this very gracious side."

FIGURE 9. Gerard P. Kuiper

Despite Kuiper's lack of interest in teaching, the students who studied under him thrived in the new field of planetary science. Toby Owen, whose distinguished career launched from his original work in spectroscopy in Tucson, said that "in all this time I never forgot Kuiper and LPL. Besides being a great scientist, Kuiper was an exceptionally kind person. I have had a career filled with marvels, and I owe that career to him."

In a time when few scientists turned their eyes toward the Moon, Kuiper's vision and genius helped forge a new field. "He had to be very strong," George Coyne said. "He was starting a major effort here at the University of Arizona, so he had to dominate the scene." Through his own formidable reputation, Kuiper returned the study of the solar system to the realm of respectable science. He brought Arizona to the forefront of planetary research, and he had impeccable timing. LPL's stature—and with it, the field of planetary science—grew with each year that brought America closer to its goal of reaching the Moon.

LOOKING BACK

In the autumn of 1946, a V-2 rocket armed with a 35-millimeter camera launched from White Sands, New Mexico. Minutes later it slammed back into the surface of the Earth. The smashed shell harbored photographic negatives, snapped off from the startling altitude of sixty-five miles. Stitched together, the black-and-white pictures showed the curvature of the Earth against a dark backdrop of space. It was the first glimpse humans had of their home.

William Hartmann was still a student at Pennsylvania State when the *New York Times* printed those photographs. Their artistry caught his eye. He was already well on his way toward a career that would merge scientific papers with paintings. Today his studio looks like a boy's playhouse. Bookshelves and easels clutter the room, which perches at the top of a weathered flight of stairs. An original Chesley Bonestell hangs on the wall: geysers spray from a cratered surface, and the flat geometry of Saturn's rings dominates the sky.

Before the Apollo era, scientists couldn't imagine the minute details of other planets with any accuracy. Humans could scarcely picture their own little world. Drawings in children's books depicted the Earth as a fully illuminated sphere, hanging like a Christmas ornament against a deep blue sky with the sharp edges of Africa, Europe, and the Americas proudly delineated. No one knew what the Earth actually looked like from a distance.

Space artists like Bonestell seized on the V-2 photographs for inspiration, unaware that New Mexico had a very specific kind of cloud pattern. "There are lots of these little individual cumulus clouds, and they would actually cast a shadow," Hartmann said. "Bonestell would paint the Earth this way, with these little patchy clouds. Nobody realized that these clouds were organized into these huge systems, these big cyclonic bands and spirals. People knew a hurricane was a spiral, but the early artists trying to understand what the Earth would look like from space didn't sense the extent of it. They painted all the clouds as separate, because that's what you could see from the V-2 photographs in New Mexico."

In August 1966, Lunar Orbiter 1, dipping low over the Moon's surface, sent back a black-and-white photo of the crescent Earth above the cratered surface of the Moon. The Earth appeared brittle as glass against the darkness of space. The photo caught scientists by surprise. You could barely see the continents. They disappeared beneath huge, shifting systems of clouds.

The photos of Earth from the Lunar Orbiter missions were unplanned and went largely unnoticed by the public. The picture from the crew of Apollo 8, however, was in color. Half shrouded in darkness, Earth drifted alongside the pale bulk of the Moon. It was just before Christmas. Frank Borman, James Lovell, and William Anders would spend the holiday reading from Genesis as they circled the Moon in their metal capsule.

That photograph shocked the world. "Here was this little blue marble sitting there in black space, and you could hardly see the atmosphere," Robert Strom remembered. "Then I think it dawned on people, wow, we live in a precarious environment, and the only thing separating us from death is this thin atmosphere of oxygen and nitrogen."

It was the beginning of a shattering transition into an age when images would resolve our unimaginable solar system into tangible worlds, each one a vastly different and haunting mirror of our own small planet. Humans could at last picture the place beneath their feet, all its vastness contained into a single, crystalline drop of blue. Fences, boundaries, and civilizations washed away beneath an immensity of cloud and water. For the first time, we recognized home.

PART THREE

TEACHING

DESPITE THE EXCITEMENT of the era, planetary science had not quite earned its place among respectable institutions of astronomy. When an expert on planetary atmospheres named William Hubbard took a faculty position at LPL in 1972, his colleagues looked at him askance. "In those days it was dismissed by many astronomers as the Loony Lab," Hubbard said, "a place where you had rather eccentric people who were under the sway of a dictator, namely Gerard Kuiper, who was not particularly enlightened in his approach to things."

LPL expanded rapidly in the fervor of the Space Race, but money from NASA, Kuiper knew, would not last forever. The lab's endurance depended on its ability to draw funding from other sources. With the university's encouragement, he intended to transform the research lab into a teaching department, offering advanced degrees in the new field of planetary science.

Hubbard recalled,

[Kuiper] was a very energetic person, especially given his age. He was very enthusiastic about his new department, and he took me on a tour of all of his observing sites around the area. He talked to me about where he thought the laboratory was heading and what he thought my role would be in it. The way he expressed it to me was that the department was going to be an essential component for keeping the laboratory in existence. He thought that in order to ensure the longevity of the whole enterprise that we needed an academic arm, we needed to have graduate students, we needed to have a teaching program.

Although Kuiper's assistants often lamented his lack of interest in their education, Kuiper had always intended to partner the Lunar Lab with a teaching department. As early as 1962, he requested information from his students about their classes—especially the off-beat geology work guided by Spencer Titley—with the goal of shaping a program in "planetology."

The idea garnered support from high places. A young chemist named John P. Schaefer succeeded Harvill as president of the University of Arizona in 1971. Schaefer intended to establish the university as a prominent research university. Tucson's climate, geography, and history had fostered interdisciplinary programs in a broad range of sciences. Schaefer recognized that excellent research units like the Laboratory of Tree-Ring Research, Steward Observatory, and LPL could attract federal funding to support nonteaching researchers, necessary to augment state-funded teaching positions.

Like Harvill before him, Schaefer had ambitious aspirations for LPL. The creation of an academic department was essential. Faculty positions were scarce in the space sciences. Most planetary scientists built their careers at NASA laboratories and aerospace corporations. A department would draw steady funding from the state and give LPL researchers the legitimacy of an academic title. It also tied the somewhat independent research organization into the traditional structure of the university, giving LPL plenty of room to grow.

Steward Observatory's new director, Ray Weymann, also endorsed the idea. In 1970, Weymann inherited from Bart Bok an astronomy department distorted by the increasing demand for solar system studies. As more and more students arrived at the university with a passion for planets, the Department of Astronomy expanded its class offerings to accommodate them.

Yet the two fields had become very different. Astronomers studied the high-temperature physics of exploding supernovae and nascent galaxies while planetary sciences required a physics of cool bodies—planets and moons warmed by the Sun. Astronomers had little need for geology and atmospheric sciences, components that were vital to the planetary field.

Pat Roemer, always a champion for students' interests, soon became concerned about the suffering academic performance of the planet-focused scholars. "The astronomy department at the present time is accommodating a number of students whose interests are orientated more toward planetary science than toward stellar astronomy," she wrote in a memo to Kuiper. "It is apparent that some very able students find the going not entirely suited to their interests."

Recognizing the need to refocus his department on astronomy, Weymann supported the proposal to begin a Department of Planetary Sciences that could provide a home for the troublesome students (as had Bok before him). The university's executive vice president, Albert Weaver, and the vice president for research, A. Richard Kassander Jr., both had training as physicists. Along with Schaefer and Vice President for Finance Gary Munsinger, they worked most Saturday mornings to make the new department a reality.

In April 1972, the College of Liberal Arts approved the formation of a Department of Planetary Sciences. Roemer headed the committee appointed to design the new department, which included Kuiper and Frank Low from LPL, Bok from Astronomy, Leon Blitzer from Physics, and Louis Battan from Atmospheric Sciences. The committee arranged interviews with prospective faculty, designed the new department's budget, decided on degree requirements, and even began to teach classes. Prior to the department's official inauguration, they arranged a series of guest lectures to offer classes or colloquiums on the newest research.

"Those were electric times, I must say, to have all these new people come in who were specialists in fields outside of planetary astronomy," Steve Larson said. "There were geochemists; there were plasma physicists; there were experts in the formation of the solar system. Every big name in the field was coming. It was just a tremendous time of growth."

A NEW DIRECTOR

The Department of Planetary Sciences officially began on July 1, 1973, not long after Kuiper's retirement as director. Kuiper remained active in teaching and research, but Charles Sonett, Kuiper's handpicked successor, would take on a dual role as LPL's director and first department head. Sonett arrived at the Lunar Lab in the midst of its transition from research organization to teaching institution. As Schaefer worked to create state-funded positions at LPL, Sonett brought his own connections and experience to draw federal funding.

Sonett grew up under California's dry desert skies. His interest in amateur astronomy and arid land geology soon developed into fascination with the design and construction of scientific apparatus. World War II disrupted his study of physics, and he lost his left leg to a German land mine in France in 1945. He returned to California to complete his spotty undergraduate studies and talk his way into a graduate program. He received his PhD in 1954 just in time to enter the nascent Space Race.

Sonett was working for Space Technology Laboratories, a civilian scientist among Air Force engineers, when the Soviet Union launched Sputnik. All at once, the company's theoretical work had to become the hardware that would take America into space. "I happened to be in the right place at the right time," Sonett said. He received the task of designing the payload for the Pioneer missions, and in 1960 transferred to NASA headquarters in Washington, DC, to become involved in the planning stages of Apollo. There, he made an initial commitment of funds to Kuiper's upstart planetary program in Tucson.

In 1962, NASA created the Space Sciences Division at the Ames Research Center in California and appointed Sonett as its leader. Despite its small size, the new division plunged into the highly competitive world of instrument building. A physicist by training, Sonett championed for greater scientific involvement in the Apollo missions, campaigning for instruments to study phenomena like solar winds and magnetic fields.

"It was a very intense time for people working in space," Sonett said. "Spent a lot of time in the lab—I remember sixty-, eighty-hour weeks. If you're getting ready for a flight, you know, you don't have time to sit around. You have to work day and night."

The Apollo magnetometer was his crowning achievement. "A big moment was when my instrument was unfolded and turned on," Sonett said. "It worked; we started getting data back right away." Earlier, Sonett had developed magnetometers, which measure magnetic fields, for the Pioneer and Explorer satellite programs and the Mariner 2 flyby of Venus. Apollo 12 brought Sonett's instrument to the Moon, and each subsequent lunar landing carried a magnetometer.

Astronauts carried the awkward boxes with gold-covered booms out to the lunar surface, where they unfolded like gangly flowers. The automated instruments looked for clues to the puzzling question of whether the Moon had an iron core. Decades later, an artist's elegant sketch of a magnetometer, rendered in dusty lunar grays, hung in Sonett's home.

At LPL, Sonett became the head of the truly interdisciplinary Department of Planetary Sciences. Faculty members came from a wide variety of departments, including physics, atmospheric sciences, and astronomy. Kuiper taught a course on planetary atmospheres, Tom Gehrels on polarimetry, and Roemer on comets. Donald Davis, a new hire, taught celestial mechanics, and physicist Leon Blitzer explained perturbation theory. William Hubbard, Robert Strom, and Harold Larson became the department's first associate professors. Geologist Michael Drake and cosmochemist Laurel Wilkening joined them in 1973.

"One got the impression that things were going to change very drastically here at that time," Hubbard said, "and that's one of the reasons why I was enthusiastic: because it looked like we were getting in on the ground floor of something where you could really have an impact."

FAREWELL

Kuiper's small laboratory had flourished under Tucson's scorching skies. The summits of Mt. Lemmon, Mt. Bigelow, and Tumamoc Hill glinted with telescope domes, and planning for the new department was well underway. The advent of high-flying planes and spacecraft had drawn the planets closer. Yet Kuiper never gave up his search for the perfect telescope site.

"Right after I arrived," Wilkening said, "Kuiper was invited to go to Mexico to help locate an appropriate site for an observatory." Godfrey Sill, a chemist and Carmelite friar, arranged to pilot a light airplane over the mountains for the trip. Sill had been a high school chemistry teacher in Illinois before coming to the University of Arizona for his PhD in 1965. In the early seventies, his laboratory experiments on the spectra of frozen sulfuric acid particles solved the mystery of the composition of Venus's layered yellow clouds.

At LPL, Sill shared his chemistry lab with Wilkening. He convinced Kuiper to bring her along as an interpreter. "We flew off to Mexico and went around to all of these sites, going back and forth across the Sierra Madres and staying in these little places," Wilkening said. "What I had was *Frommer's Five Dollars a Day Guide to Mexico*. Kuiper was a notoriously tight-fisted, penny-pinching guy, and he just loved these seedy hotels that we stayed in."

Shortly before Christmas that year, Kuiper returned to Mexico City with his wife Sarah and longtime friend and fellow astronomer Fred Whipple. Then, on December 24, 1973, Kuiper died of a heart attack in his hotel room at the age of 68. Reporting his death on Christmas Day, Tucson newspapers praised his bold interpretations of astronomical observations. He had only recently announced his discovery that Jupiter's Great Red Spot was an anvil-shaped permanent storm. His cramped writing fills the calendar pages of his planner in the months before his death, noting faculty meetings and lectures. He remained committed to seeking new knowledge until the end.

Kuiper's life entered the realm of legend in the halls of LPL. Two years after his death, the C-141 aircraft was renamed the Gerard P. Kuiper Airborne Observatory (KAO) in recognition of his efforts to direct astronomy in this new direction. Kuiper had walked through the echoing cabin of the plane before its maiden flight, but he did not live to see it lift off from the ground.

Kuiper's name also graces the belt of icy objects slinging around the solar system beyond Neptune's orbit, which Kuiper, Whipple, and other astronomers had hypothesized for decades. Astronomers discovered the first objects in the belt with the telescope on Mauna Kea that Kuiper worked so hard to establish. Kuiper is honored with craters on Mars, Mercury, and the beloved Moon that he devoted himself to mapping. For many LPL scientists, his presence lingers in the odd corners and brick passageways of the Space Sciences Building that now also bears his name.

Kuiper's ashes, and those of his wife Sarah, are interred at an unmarked place near the 61-inch telescope that was his pride and joy. Astronomers are, above all, romantics, and it is easy to imagine Kuiper pacing the big dome building at night, his footsteps muffled by the groaning of gears as the telescope swings into position.

GROWING PAINS

The first students to pass through the new department were young, lively, and tightly knit. Just as Titley's geology students had been children of Sputnik, these were children of Apollo, inspired by the Space Race and staking their careers on its continuation. "It was a brand-new department, with no history, with no traditions," said Guy Consolmagno, one of the early graduate students. "We were inventing the traditions. We were inventing our own history."

The students received guidance from a faculty that was also young and lively, and in some cases completely unprepared to teach. Wilkening, an expert in geochemistry and cosmochemistry, recalled her consternation at teaching those subjects to classes filled with math and physics majors. "It was challenging to have students who were very bright and had very strong backgrounds," she said. "We were eclectic because there weren't many places that students could come, and there was a lot of interest in the space program."

Uwe Fink took over Kuiper's class on planetary atmospheres and struggled, like many of his colleagues, to balance his first love—research—with his new duties. "Our lectures were sort of like Kuiper's—we just told them what we were doing at the telescope," he said. "The good students really liked it because they got first-hand experience of how research works. But other students expected canned stuff, like they were getting in their other classes."

Over time, faculty members learned how to communicate with their students, and with each other, despite the specialized jargon of each field. The students, meanwhile, suffered through the department's growing pains in a trial by fire. Hubbard taught a rigorous course on the interiors of planets, Strom guided students through the surfaces of rocky bodies, and Hartmann lectured on the origin of the solar system. The teachers gave exams they admitted they couldn't pass themselves. All specialists in their own fields, none of them had been educated in the whole breadth of planetary science.

Invigorated by the success of Apollo, NASA launched missions to Mercury, Venus, and Mars, gathering unprecedented views of their evocative landscapes. It was a new era in space exploration, and Sonett seized the opportunity. LPL began shifting its focus toward spacecraft and theoretical research, deepening its distinction from astronomy departments that remained reliant on ground-based telescopes.

The transition did not occur seamlessly, inciting unpleasant schisms between the "old guard" planetary astronomers and newly hired faculty. In 1977, the telescopes in the Catalina Mountains were consolidated under the management of a single entity, the University of Arizona Observatories. Gradually, ground-based research became the domain of Steward Observatory, and several researchers transferred to the Department of Astronomy while others held joint appointments with the Lunar Lab. Fruitful collaborations took place across the two institutions, but the gap between planetary science and astronomy continued to grow.

Sonett hired new researchers to analyze the last dregs of data from the Apollo program. One of them, Floyd Herbert, had long been a hanger-on at the Lunar Lab, ever since rooming with Chuck Wood years before. Another, Lonnie Hood, had been paying for his college tuition by sandpapering school buses in the summer of 1969 when he took a week off to see Apollo 11 launch from Cape Canaveral. Now working in Sonett's lab, Herbert and Hood analyzed data from the very missions that had inspired them to study space.

Everyone had hoped that Apollo samples would settle the longstanding issue of the Moon's origin. Instead, they sparked more questions. The lunar crust had startling similarities to Earth's mantle, yet it contained very little iron and no water or volatile materials. The Moon appeared to lack a metallic core or have only a very small one. And, while lunar rocks held records of substantial past magnetism, the modern Moon had no global magnetic field.

Hood and Herbert, working with data from Sonett's magnetometer, built a case for an iron-rich core with a radius no more than 435 kilometers by comparing measurements of Moon's fluctuating magnetic fields taken from the surface to those taken from high orbit. Any model attempting to explain the Moon's origin had to account for this relative lack of iron. In short, the Moon appeared to be neither stranger nor sibling: it was too similar to Earth to have been captured from somewhere else, and too different to have grown alongside it.

In 1974, William Hartmann and Don Davis—recent cofounders of a new group in Tucson, the Planetary Science Institute (PSI)—decided to revive the languishing question of the Moon's origin. Davis, who had a PhD in physics from the University of Arizona, had returned to Tucson after developing computer systems for Apollo. He had garnered fame as a member of the team that safely returned the astronauts of unlucky Apollo 13 back to Earth. Davis taught a course on celestial mechanics at LPL before founding PSI with Hartmann.

They already knew, from the clues left by the Orientale Basin and other bull's-eye structures, that the Moon had suffered impacts from large objects. As a graduate student, Hartmann had the leisure to read the work of Soviet scientists, and he became familiar with Victor Safronov's theories on planetary accretion. Safronov pointed out that if planets spun into existence from countless small bodies called planetesimals, it was likely that a few competing proto-planets appeared in the same vicinity. Therefore a forming planet could suffer not just many small impacts, but a few large collisions as well.

Hartmann and Davis calculated that during Earth's formation, potential competitors might have grown to one-fourth or one-half of Earth's size. They proposed that a massive collision could have blown Earth's mantle into space, forming a dust cloud of Earth-like material—but lacking the iron that had already sunk to the core—that later aggregated into the Moon.

When Hartmann presented the idea in a lecture at Cornell University, fresh from his graduate studies, he braced himself for critique. He didn't expect a respected Harvard astrophysicist, A. G. W. Cameron, to raise his hand and announce that he had been working on the same theory with a colleague, William Ward. Cameron had one correction to make: his calculations showed the colliding object might have had the mass of Mars.

Hartmann and Davis, and Cameron and Ward, published their results separately in 1975 and 1976. The scientific community proved reluctant to accept the giant impact theory, which seemed ludicrously random. Hartmann, however, felt encouraged by the long odds. The Earth had a Moon like no other in the solar system. Other planets had tiny, odd-shaped moons, or whole systems of moons—like miniature solar systems. Only Earth, out of eight or nine planets, had suffered the catastrophic impact that would create such an enormous moon.

The question remained open for another decade. It didn't help that in a post-Apollo slump the world once again lost interest in the Moon. Spacecraft missions revealed newer and more fascinating worlds to explore. During the sixties, NASA sent four Pioneer missions into solar orbits, demonstrating the feasibility of spinning spacecraft. They were now in the midst of the ambitious Pioneer 10 and 11 missions to Jupiter and Saturn.

Also in the sixties, NASA launched eight Mariner missions, five of which reached their destinations—two to Venus, three to Mars. When Mariner 9 (the first spacecraft to orbit another planet, rather than simply fly by) arrived at Mars in November 1971, it found the planet swathed in a global dust storm. The scientists had to wait until December for the dust to subside, revealing lofty volcanoes, jagged chasms, and insubstantial morning fogs.

NASA's next spacecraft, Mariner 10, slung around Venus in February 1974 and arrived at Mercury in March. The eventful journey, during which flaking paint repeatedly distracted the spacecraft from its guide star, marked the first use of a gravity assist. Widely used on later missions, gravity assists made operations cheaper by allowing the spacecraft to fall toward a planet and gain speed before slingshotting out again. Mariner 10 discovered no notable atmosphere around Mercury's heavily cratered, Moon-like surface. The interior, however, had more in common with Earth, marked by a magnetic field and an iron-rich core as large as Earth's Moon.

The mission sparked a lifelong passion for Mercury in Robert Strom, who served on the science team and would not have another chance for an up-close look for three decades. Kuiper had also contributed to the mission's planning but did not live to see Mariner 10 arrive.

Out of this wealth of spacecraft data, new specialties began to emerge. "You can build a laboratory in many ways," Sonett said. "The way we chose was to try to build the laboratory with a broad base of scientists that had different capabilities." Chemistry, geology, and atmospheric physics would be essential to the field in a way they hadn't been to astronomical research. With encouragement

from Schaefer, Sonett expanded the faculty, inviting solar physicist Randy Jokipii and astrophysicist Eugene Levy to join the laboratory. The study of energetic particles and solar wind barely existed in science departments, and Sonett
was determined to carve out a niche for it at LPL.

"Space science is a funny business," he said, "because it covers a lot of disciplines of science, and so you almost have to be a master of many different trades
to work in it."

Sonett also made a home for the flood of new photographic data. He nego-
tiated with the university administration and NASA to share the costs of the
new Space Imagery Center for image storage and processing. The university
provided space in the Kuiper Building and money for the renovation, while
NASA agreed to pay for equipment and the salary of a full-time librarian. The
Space Imagery Center opened in July 1977, the first of its kind in the country.
Pleased with its success, NASA now has seventeen similar centers around the
world. More than just repositories for data, these facilities assist researchers,
educators, the media, and the public.

LPL was well on its way toward becoming a research institution on par with
JPL or any of NASA's research centers. "As I was told by the executive vice
president of the university," Sonett said, "they don't want just another laboratory. It has to be the best."

IMAGING JUPITER

Martin Tomasko built his own telescope as a kid, devouring books on the night
sky and even grinding his own mirrors. But tinkering with machine bolts and
clock drives didn't take him into astronomy. Instead, Tomasko started his freshman year in college studying mechanical engineering, convinced he couldn't
make a living by looking at stars.

After spending a summer at the drafting table, watching the shadow of the
telephone pole outside his window tick off the dull hours, Tomasko changed
his mind. It was the sixties, when people like Wernher von Braun and Gus
Grissom gave inspirational speeches to students around the nation. "I thought,
the space program is going great guns," Tomasko said. "If there's ever a time to
try to make a go of it in astronomy and space science, this might be the time."

Riding on the success of the balloon program, Tom Gehrels had written an
eight-hundred-page proposal to NASA to build an imaging photopolarimeter

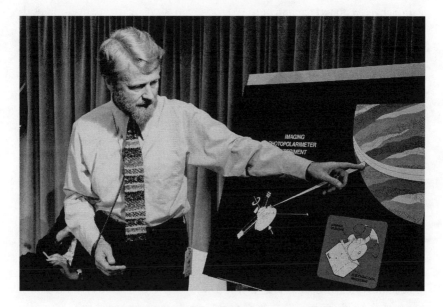

FIGURE 10. Tom Gehrels speaks at a press conference
during the Pioneer mission in November 1973.

COURTESY OF NASA AMES RESEARCH CENTER

for the upcoming Pioneer 10 mission. At first NASA hadn't planned to capture
photos with Pioneer 10. Gehrels wanted to study Jupiter's atmosphere, but he
realized his instrument could accomplish imaging as well by taking advantage
of the spacecraft's spin—a new technique called "spin-scan imaging."

When Tomasko arrived at LPL in 1971 with a PhD in astrophysical sciences from Princeton, he jumped on the opportunity to join Gehrels's team.
The instrument, after all, would return the world's first close glimpse of Jupiter
just two years later. Although ground-based telescopes could easily photograph
bright Jupiter, Pioneer 10 had one great advantage: it could fly all the way
around the planet. "We could observe Jupiter in all these different new geome-
tries that we'd never seen before," Tomasko said, "and we could learn something
about the nature of the cloud particles by flying around and looking at it from
other directions."

The photopolarimeter didn't work like a conventional camera. Made of
lightweight magnesium, it measured light from a single point in the red and
blue wavelengths as the spacecraft spun, building up an image one small sliver at

a time. Later a converter system would add green to create "cosmetically appealing" images for TV. The spacecraft itself spun continually about the axis of its radio dish, which pointed toward Earth. It communicated through the impressive Deep Space Network, three complexes of receiving systems and dish antennas placed strategically in Australia, Spain, and California's Mojave Desert.

The tiny observatory, its innards a maze of detectors and electronics, did not sit idle during the cruise. Gehrels included a wide aperture on the telescope to study the light scattered from interplanetary particles, which produces a faint glow spanning Earth's sky along the zodiac at sunset and sunrise. It was a risky move—bright Jupiter would ruin the instrument if something went wrong and the wide aperture couldn't be closed—but it made good use of the long journey.

An astronomy student, Lyn Doose, joined the team just in time for Pioneer 10's closest encounter with Jupiter on December 3, 1973. The LPL researchers traveled to the NASA Ames Research Center in California to see the first data return in real time. "It wasn't like today where you have these vast computer memories on spacecraft," Doose said. "Basically the instrument couldn't remember anything but what it was told last. There were tens of thousands of commands, and everything was on IBM punch-cards."

A tense moment occurred early in the encounter when the photopolarimeter, scrambled by radiation, began looking for the zodiacal light with its wide-open aperture. If the spacecraft turned to scan Jupiter, brilliant light pouring through the opening would destroy the sensitive detectors. Wild signals arrived from other instruments as well, and for a few despairing hours it seemed that Jupiter's powerful radiation belts—created by the planet's magnetic field trapping charged particles from the solar wind—would bring the mission to an inglorious end.

Soon, however, clear signals began to arrive. Spinning majestically, the spacecraft swept out lines of photometry across the surface of Jupiter. The photopolarimeter captured scattered sunlight from Jupiter's swirling clouds and converted the data into shades of red and blue. The angle of the telescope could be stepped in half-millimeters, adjusting each scan line slightly. "All the while that it is building up an image, it's getting closer and closer to Jupiter," Tomasko said. "The image is not like a snapshot. It's accumulated over an hour and a half, and it's got ferocious geometrical distortion."

The scientists at the NASA Ames Research Center separated the imaging data from the rest and sent it to Tucson on a dedicated telephone line. Later, the data would arrive on reels of magnetic tape for more detailed image processing. The Optical Sciences Center had developed an impressive computer system to assemble the data, extrapolate the green color, and provide real-time photos to the press, a feat that won the Pioneer program an Emmy award. It also had to remove distortion from the spacecraft's spin, which made Jupiter appear oddly stretched. To the chagrin of the research group, the press distilled their attempts to explain the advanced techniques into the unfortunate statement that Jupiter would look like a banana.

As Pioneer 10 drew closer, Jupiter's radiation belts worsened the distortion. Occasionally, energetic particles whacked the photopolarimeter, causing it to skip randomly over the planet's surface instead of scanning uniform lines. Roundtrip communication time spanned ninety-two minutes, so by the time mission engineers at Ames received information that the spacecraft hadn't correctly followed their instructions, they had already sent forty-six minutes of new commands.

"Commanding the instrument was like telling an old senile guy to go to the store to get a quart of milk," Tomasko said. "The instrument would start moving the way you wanted, and then halfway through it would just forget its mind."

After the encounter with Jupiter, Pioneer 10 headed for the outer solar system, destined to cross Pluto's orbit almost two decades later and sail onward into space. Pioneer 11, its sister ship, drew close to Jupiter a year later, also bearing an imaging photopolarimeter. It survived the nail-biting crossing through the radiation belts, skipping low over the turbulent cloud tops. Doose programmed the sequence that photographed the famous Great Red Spot, an image that appeared on the cover of *Scientific American*. Along with two other University of Arizona students, Cliff Stoll and A. E. Clements, Doose completed his PhD dissertation under the Pioneer program.

For many scientists, a new awareness of the planet's composition eclipsed even the most stunning images. They could now translate bright bands and dark polar regions into intricate interactions within Jupiter's atmosphere. Brimming with excitement, LPL researchers pored over spectroscopy and photopolarimetry data that revealed layered clouds packed tight with hydrogen, helium, and a haze of ammonia. Meanwhile, 500 million miles away, Pioneer 11 swung around Jupiter and set a course for Saturn.

HAWTHORNE HOUSE

In 1976, one of the graduate students found a house for rent on Hawthorne Street, a few blocks east of campus. For most students, their modest stipend made affording a house out of the question, but this one had five rooms. Guy Consolmagno, Bob Howell, John Wacker, Mike Feierberg, and Bruce Wilking made up the original residents of what came to be known as "Hawthorne House."

When Howell, a bicycling enthusiast who wanted to live farther from campus, moved out, he was quickly replaced by Nick Gautier, a student in the Department of Astronomy. Graduate students continuously occupied the house until its owner ceased renting it in 2005. It was the social hub of the Department of Planetary Sciences, and students could always depend on having a corner in Hawthorne House when they didn't want to go home.

A motley collection of rebels and pranksters made up the first residents of Hawthorne House. Living and working in close quarters, they played endless practical jokes. A measure of their creativity is the T-shirt printing company, whimsically named "Nocturnal Aviation," instigated when John Grady and Cliff Stoll designed shirts with the Lunar Lab's emblem. Nick Gautier wrote letters to companies offering them custom T-shirts in exchange for lab equipment. Boxes of surplus apparatus and old computer terminals began appearing at the lab, and once, a "thing on wheels" the size of a telephone booth that turned out to be a high-voltage source for vaporizing aluminum. Texas Instruments offered the students four sets of chips for mini-computers in exchange for five thousand T-shirts. The students rose to the challenge, and years before Apple or IBM, they built their own computers on the kitchen table at Hawthorne House.

The rocket-fueled wonder of the 1960s gave way to even more fantastic dreamscapes. In 1977, the movie *Star Wars* swept the box office, pushing the limits of what humans believed was possible both in filmmaking and in space. Students crowded into John Wacker's green Volvo and headed to Edwards Air Force Base to see the shuttle Enterprise land—"like flying a brick," Consolmagno said—and then continued to Los Angeles, which had one of three theaters in the world showing *Star Wars* in seventy millimeters.

Like LPL's original students, they were a close bunch, united in a common struggle to define their chosen field. In elementary school, they had drawn inspiration from satellites tracing arcs through the sky. Consolmagno had bought his first telescope with grocery store trading stamps. "Everybody thought space was the future and little boys should be scientists," he said.

Mostly mentored by astronomers, the students broke free from traditional telescopic methods and adapted to rapidly evolving technology. Nicholas Schneider, newly arrived at the Lunar Lab in 1979, recalled his astonishment to find an entire building dedicated to the study of planets. "There were very few

FIGURE 11. A rare Tucson snow falling on Hawthorne House, January 1987

COURTESY OF JOHN SPENCER

places in the world where you could do planetary science," he said. "When I was applying for graduate school, my astronomy advisors—because there were no planetary advisors—said, 'Nick, you're giving in to the Dark Side,' a term which was two or three years old. I'm very proud of it, and I feel very lucky that the timing worked out to realize that this was the wave of the future."

PROTÉGÉS

Fifteen students left the Lunar Lab with hybrid degrees before the Department of Planetary Sciences accepted its first generation of official graduates in 1973. While working at LPL, these early students obtained master's degrees and doctorates in the Departments of Astronomy, Geology, and—in the case of Sam Pellicori—Optical Sciences.

Toby Owen, a few years ahead of the others, had earned his PhD in the Department of Astronomy in 1965, writing a dissertation on planetary atmospheres under Kuiper's tutelage. He joined a research group at the Illinois Institute of Technology called IITRI. Binder followed him there after earning his PhD in geology, but the newly formed space group, focused on developing post-Apollo missions to the Moon and planets, found it difficult to attract planetary scientists to cloudy Chicago. At Binder's urging, IITRI opened a Tucson branch.

By 1971, the group included several young scientists with ties to LPL: Binder, Hartmann, Davis, and Clark Chapman, a physicist from the Massachusetts Institute of Technology (MIT) who had spent summers in the early sixties measuring craters for Kuiper's lunar effort. That group became the core of PSI, which started out as a desk in Hartmann's living room in early 1972. Members of the fledging nonprofit soon became known for their work on lunar and Martian cratering, asteroids, major impact theory, and the origin of the solar system. Tucson now had yet another foothold in the nation's fast-moving ascent into planetary research.

Dale Cruikshank earned a PhD in geology in 1968 and spent a year studying in the Soviet Union. Determined to learn the secrets of science behind the Iron Curtain, he had studied the Russian language intensively in college, a highly unusual choice at the time. With Kuiper's help, he enrolled in an exchange program to learn from parallel efforts in infrared astronomy. He returned briefly to LPL before deciding to strike out on his own, transferring to the newly prominent astronomy department at the University of Hawai'i.

After a stint in the Peace Corps, Chuck Wood returned to LPL for his master's degree and then accepted a university teaching position in Ethiopia, where he studied the dramatic rift faults cut into the landscape. He returned to the United States two years later to continue studying planetary science. Times had changed. Kuiper, Urey, and Whipple were no longer the only experts for students to emulate. Wood earned a PhD in planetary science from Brown University and found many opportunities to follow his avocation.

"One of the things I realized, somewhere along the way, is that I was really lucky," Wood said. "Having scientifically grown up at the Lunar Lab—which was the first planetary science group pretty much in the country—and because of not staying in one place, I ended up getting the chance to meet a lot of people who were important scientists for studying the solar system."

LPL's protégés were among the few people prepared for the nation's grand venture into space. To his surprise, Hartmann received a phone call from Bruce Murray, JPL's director, inviting him to join the imaging team on the Mariner 9 mission to Mars. "It falls right into my lap," he said. "I contrast that with today:

when a new mission gets announced they'll be two hundred bright, bushy-tailed scientists with fresh PhDs trying to get on. . . . It was a great time to be doing this stuff because there weren't very many young people coming out with degrees."

Binder became a principal investigator of the imaging team for Viking 1, the first lander to successfully touch down on Mars. Early in the morning on July 20, 1976, Binder awaited the images in breathless anticipation. "It was a slow-scan camera," Binder said. "Nothing like what you see now where the pictures just come down, boom, boom, boom. It was a facsimile camera, so it did one scan at a time in the vertical and slowly it would rotate."

The photographs rendered Binder speechless under the glaring lights of TV cameras. "My God, this panorama began to come in front of me, and there were rocks all over the place, and I didn't know what to say," he recalled. "I was just dumbfounded by the beauty."

No one really knew what the first images of Mars would reveal: it was possible to imagine moss-green boulders with blue skies above. Kuiper himself had popularized the idea of lichen on Mars. Even when the rust-red desert unrolled before the cameras, void of visible life, scientists could hold out hope for microbes invisibly thriving in the soil.

Owen had one of the coveted slots on the soil analysis team, which paired two simple but incredibly precise instruments—a gas chromatograph and a mass spectrometer—to measure the masses of molecules and atoms. From samples spooned into the instruments, scientists could infer what compounds were present on Mars. They built a picture of a volcanic terrain, salty and sulfurous, irradiated by ultraviolet rays and shrouded in a thin carbon dioxide atmosphere—a hostile place, and to all appearances devoid of the organic compounds that accompany life.

The planet appeared utterly alien from Earth—yet strangely similar, too, with water-carved gullies and wind-shaped dunes not unlike those of the Sonoran Desert. Viking 1 and other successful missions attracted eager young students to Tucson and the burgeoning field of planetary science. In 1977, William Hubbard succeeded Sonett as LPL's director and department head. A high school senior when the Soviet Union launched Sputnik, Hubbard had been dismayed to find America lagging behind in the Space Race. After getting his physics degree, he began casting around for an appropriate graduate school.

"There was nothing like the Planetary Sciences Department," Hubbard said. "There was no direct path into the space program." He enrolled in the

University of California, Berkeley, attracted by its highly traditional departments of astronomy and astrophysics.

"My advisors at Berkeley were rather contemptuous of planetary science," Hubbard said. "They felt it was a real backwater; not real astronomy. The attitude in those days was that planetary astronomy was an area in which rather pedestrian physics was being applied, and not much was coming out of the field that was new or exciting."

That changed during Hubbard's years at LPL as spacecraft began redefining the way scientists thought about and studied the solar system. In his role as director, Hubbard eased the Lunar Lab's tumultuous transition from planetary astronomy into the rapidly changing field of planetary science. The ground-based research of Kuiper's time gave way to much broader areas of inquiry, and theoretical and spacecraft work found a place alongside of telescopic astronomy. Planets were no longer simply wandering stars. They had become worlds that could be touched and tested, solid ground where robots could stand with ungraceful legs.

ILLUMINATING ICE

Kuiper's death and the departure of his protégés left a gap at LPL in the spectroscopy program. Harold Larson continued the airborne work that Kuiper had begun while Uwe Fink agreed to take over the laboratory and ground-based projects.

As Kuiper had predicted, spectroscopy remained queen even in this age of spaceflight. Sample return missions from objects more distant than the Moon remained a dream for the future, and while landers had touched down on Mars and Venus, the outer planets still told their secrets mostly through the medium of light. The opening of the infrared region of the spectrum meant a completely new way of exploring worlds, and instrument scientists at LPL worked continually to refine the detectors pioneered there by Johnson and Low.

"Some scientific techniques run up against natural obstacles, and after you find so much, that's it," Fink said. "You can't go any further; you have to do something totally different." With infrared detectors leading to revelations impossible to make in the visible wavelengths, data were collected and left lying in archives as researchers jumped to the next object of interest—"data that people today would think were really great discoveries," Fink said.

During the seventies, Fink and Godfrey Sill embarked on a comprehensive study of ices using Fourier transform spectroscopy. Most frozen gases appear white in visible light, rendering this region of the electromagnetic spectrum useless for identification. In the infrared, however, frosts and thin films of ice leave distinctive absorption lines (dark bands in the spectra where atoms or molecules absorbed particular wavelengths of the radiation passing through). Fink and Sill worked to compile the spectra of ices they expected to find commonly in the solar system, such as water, carbon dioxide, sulfur dioxide, methane, hydrogen sulfide, and ammonia.

The laboratory results would prove useful for interpreting data gleaned from telescopes and spacecraft. Fink and Larson identified carbon dioxide frosting the polar ice caps of Mars, just before Mariner 9 arrived to confirm the discovery. They went on to measure water ice—"the most important ice in the solar system"—on the Galilean satellites (the four moons of Jupiter discovered by Galileo). Initial results showed strong signals of water ice on Europa and Ganymede. They decided to wait to publish the data until they could obtain clearer spectra during Jupiter's opposition in 1972, an ideal time to do science, when the planet is nearly fully illuminated by the Sun, close to the Earth, and visible almost all night.

"We were partially scooped," Fink said ruefully. A planetary sciences group at MIT had also taken advantage of the opposition and published their data first. Fink and Larson confirmed that the Galilean satellites showed unexpected diversity. Europa and Ganymede clearly had large amounts of water ice, Callisto had a fainter signal, and Io had no signs of water at all, though absorption bands in its spectra suggested some other, unknown frost on its surface.

Fink and Larson turned to Saturn next, making necessary improvements to the detector to view its fainter satellites. In 1976 they announced the presence of water ice on Iapetus, Rhea, Dione, and Tethys. The same year Cruikshank and his colleagues at the University of Hawai'i found methane ice enveloping Pluto, a discovery that Fink later confirmed. The thick, clear ice made Pluto nearly as bright as freshly-fallen snow, tricking early observers into thinking the planet had a much larger size. It was the first innocent step toward Pluto's demotion to dwarf planet.

Fink and Sill published their database of ice spectra in *Comets*, a book edited by Laurel Wilkening, in 1982. "That's still the classic paper on all the ices," Fink said. "There's lots of work done on that, but the improvements are made around the edges, a little detail here, a little detail there." It was an appropriate venue because Fink would soon transfer his attention from planetary atmospheres to comets, an area of research that would dominate the rest of his career.

Comets belonged to the University of Arizona Press's new Space Sciences Series, whose general editor was Tom Gehrels. Preparing to switch his focus from polarimetry to asteroids in the seventies, Gehrels realized the need for a comprehensive textbook on both topics to close the old field and open the new one. That sparked the creation of the series, and Gehrels worked hard to stimulate discussions in the planetary science community in preparation for each new book. Many of his colleagues and former students found a place among the pages as Tucson continued to lead the way in the nascent field of planetary science.

POINTS OF LIGHT

"These were the days of the US space program at its finest," wrote Gehrels in his memoir, recalling the "eleven Great Years" of the Pioneer program. Even as Pioneer 11 swung around Jupiter and began the four-year journey to Saturn, another Pioneer project was making the first foray to a planet much closer to home. A twin mission, comprising an orbiter and a "multiprobe," arrived at Venus in December 1978.

Bright Venus, splendid in the dusk and dawn sky, had once held the glamour of a possible Earth-like world. Even after the discovery of the planet's thick carbon dioxide atmosphere, scientists postulated swamps or oceans hidden beneath the clouds. When Mariner 2 proved that the planet's surface was oven-hot—about 500°C—Americans lost interest and concentrated on Mars. The Soviets, however, had launched an immensely successful series of probes at the Morning Star, returning photographs of the pebbly surface in 1975.

When America at last returned to Venus, the orbiter revealed a young surface, oddly free from the blemishes of impact craters. The biggest probe parachuted into Venus's opaque atmosphere on December 9, trailed by three little probes in free-fall. It carried a solar flux radiometer developed by Martin Tomasko, William Wolfe of the Optical Sciences Center, and astronomy student A. E. Clements. New to the team was Peter Smith, a young optical scientist with a passion for gemstones who had accepted a temporary job at LPL earlier that year. It was just a way to pay the bills while he worked toward his American Gemology Institute degree.

The probe hurtled through stacked layers of clouds and into the clear atmosphere below before melting in the heat. "A month later we had a paper in *Science* magazine," Smith said. "A few months after that I was writing programs for the thermal balance of the Venusian atmosphere, and thought of myself as a Venusian weatherman." All thoughts of becoming a gemstone dealer vanished: "This is great! What a wonderful career, exploring planets."

Smith's planetary career had just begun. He joined Gehrels and Tomasko on the imaging photopolarimeter team, which was now gearing up for Pioneer 11's arrival at Saturn on September 1, 1979. NASA had launched the twin Voyager spacecraft two years earlier, destined for the outer solar system. They had already passed Jupiter and would head for Saturn next. NASA planned to veer Voyager 2 close to the rings to preserve a trajectory that would take it to Uranus and Neptune. True to its name, Pioneer 11 would tell whether the route was safe.

Another new arrival on the team, Robert McMillan, was fresh from his astronomy PhD at the University of Texas. He had chosen his career at the age of sixteen. "I can trace that back to an actual moment: when I got a pair of binoculars for Christmas, and I set them up on a tripod and went outside," McMillan said. In the cold Cleveland night he gazed at the bright swirl of the Orion Nebula. "It was December 25, 1966," he recalled, "when I decided to be an astronomer."

McMillan saw an advertisement for the position and joined the photopolarimeter team just a month before the Saturn encounter. The team had to monitor the uplink commands and downlink data in eight-hour shifts, twenty-four hours a day. Roundtrip communication required nearly three hours, creating an odd time distortion between members of a watch. One person monitored incoming data that the spacecraft had collected ninety minutes before, while another person sent commands that wouldn't take effect until ninety minutes later. McMillan, because he was single at the time, was awarded the graveyard shift. The late hours did nothing to diminish his enthusiasm.

"That was one of the most exciting things I ever did in my career, actually being in that big control room and watching the data come down on the consoles," he said.

The team gathered at the NASA Ames Research Center as Pioneer 11 turned its electronic eye toward Saturn. Tension rose as the clock ticked toward 9 a.m. The spacecraft was now travelling at seventy thousand miles per hour, and everyone knew it had plunged through the plane of Saturn's rings ninety minutes earlier. Whether it had safely emerged was another question.

"You have to realize at that time there wasn't much knowledge about the Saturn system," McMillan said. "Even though the Pioneer spacecraft was passing Saturn at a distance that was beyond the boundaries of the known rings, we didn't know what sort of particles would be in those rings that hadn't been visible from Earth. The issue was whether the spacecraft would be destroyed at that instant, as it passed through the ring plane."

Pioneer II tested the route for Voyager 2. If the spacecraft did not survive, then NASA would have to scrap the planned "Grand Tour" of the outer solar system. In the control room, an electronic clock displayed the Earth-received time of the spacecraft's signal. At 9:03 a.m., all eyes turned toward that clock. If the signal ceased, then the spacecraft had been lost.

"During that countdown, we—at least I—forgot that I was sitting at Ames and really felt more like I was there with the spacecraft," McMillan said. "We were the ones going through the ring plane and trying to decide if we were going to survive or not."

Pioneer II's plunge through the ring plane lasted less than a second. Receiving the good news an hour and a half later on Earth, the science team broke into relieved cheers. Data continued to pour down over the next sixteen hours. The sensitive photopolarimeter detected a thin loop of ice and stone just beyond the bright outer ring. Astonished, Gehrels and his team watched the image of the previously unknown "F Ring" appear on the overhead screen one line at a time. "Nobody had ever gotten ground-based pictures that looked that like that," Lyn Doose said.

Pioneer II safely repeated the ring crossing on the other side of the planet late that night. The spacecraft swung close to the satellite Titan—already a favorite subject of the media—and then quietly exited the stage. The Pioneer program had come to a close.

"I came away from the Saturn operations in California in a daze, having been in a wondrous world of snow and ice, of particles and fields, of satellites and jet streams," Gehrels wrote later. "I did not want to leave this fascinating world, not even to go home."

Saturn, the solar system's jewel, had revealed much of its intricate beauty to Pioneer II. Titan, however, refused to give up its secrets. Tomasko had spent a sleepless night during Pioneer II's approach planning the commands for an imaging sequence that would ensure Titan would appear in the frame. The presence of methane, detected by Kuiper in the forties, made the satellite an intriguing place to study organic molecules and look for life, but Titan's impenetrable cloud layers made it impossible to scrutinize its surface.

FIGURE 12. Peter Smith displays the newly discovered F Ring.

COURTESY OF LPL SPACE IMAGERY CENTER

It felt like a brand new way to do science: up close and personal. The Mariner, Pioneer, and Viking programs had captured the first glimpses of worlds that might have contained almost anything, from swamps to frozen oceans, from veins of coal to fields of ferns. Irreproachable fact had begun to replace imagination, but the facts were fantastic enough: water-carved Mars and cratered Mercury, the furnace burning beneath Venus's beautiful clouds, Titan's opaque mystery. The next decade would bring even closer looks at the outer planets, and a new round of glory days for space exploration.

Tomasko explained:

I came here with a degree in astronomy. These things were all space astronomy; they were little specks of light in the sky. You couldn't resolve that Titan is one arc-second across. It's just a speck of light, and the game was how much can you learn about the physical situation on this object from the light. But now of course we've been to most of these places, in orbit around them, or with entry probes going to them. They're geophysical places now. We know what they look like; we know what the ground looks like. So there's been an evolution over the last couple of decades from points of light in the sky to real places.

PART FOUR

VOYAGES

A S A CHILD, Laurel Wilkening had been dazzled by the dark skies awash with stars above her father's mountain cabin in New Mexico. She studied chemistry in college, examined lunar samples for her graduate dissertation at UC San Diego, and obtained a postdoc position at the University of Chicago, inspired by the few female professors who managed to obtain teaching positions in science departments. Strong-willed and soft-spoken, she continued to champion for women's rights throughout her professional career, demanding equal pay for equal experience.

Wilkening's good-natured personality charmed Godfrey Sill during their journey over the mountains of Mexico searching for telescope sites with Kuiper. Within a year, she and the former Carmelite friar had married. As they continued their careers at LPL, Wilkening became a strong advocate for the creation of a women's studies program, committed to attracting more girls into science and engineering careers.

"Everywhere, my whole life, there have always been more men than women, so I was used to working with guys," she said. "It didn't really bother me too much, for myself, after I left Chicago. But I realized that there were other women who were not as willing to defend themselves as I was." Through her advocacy for women's studies at the University of Arizona, she became well-acquainted with senior administrators.

In 1980, Wilkening took a leave of absence from LPL to visit NASA Headquarters in Washington, DC. She intended to join the team planning a mission to Comet Halley, which would make its perihelion passage (close to the Sun) in 1985–86. Not long after her arrival, NASA administrator Thomas Mutch died in a mountaineering accident in the Himalayas. The planning process ground to a halt, and the United States never fielded a mission to Halley.

The eighties marked a difficult decade for American space exploration. The twin Voyager spacecraft, launched in 1977, made their first approach to Jupiter two years later and transmitted a treasure-trove of data on the outer planets for the rest of the decade. But America sent no new spacecraft to follow in their wakes. Although NASA had plans underway for a follow-up mission to Jupiter and another one to Venus, a twelve-year gap would occur before Cape Canaveral saw another interplanetary launch. Meanwhile, the elaborate Voyagers had to be continually rescued from funding shortfalls that threatened to cut the Grand Tour short.

Voyager 1 made its close approach to Saturn on Veteran's Day, November 1980. At NASA Headquarters, Wilkening stopped by her office in jeans and a sweater to catch the live feed. Then she heard a knock on the door. Her presence was requested at the White House.

The summons took Wilkening by surprise. Jimmy Carter had just lost the election to Reagan, and she had forgotten that the president had named her the science representative for the Voyager encounter. She took a cab back to her apartment to change into a suit.

"I sat between President Carter and Rosalyn, and the pictures came in," Wilkening said. She couldn't help thinking of the first time she'd seen Saturn though a telescope as a young girl, amazed to see the tiny, delicate rings. Now she gazed at great loops of stone and ice, creamy gold against stark black, so stunning in their detail they seemed near enough to grasp.

The TV stations played the footage over and over again in the following weeks. "That Voyager view of Saturn, which showed the incredible richness of the rings and the structure—it was so dramatic and it was so gorgeous," Wilkening said.

She recalled later, with a laugh, that the politicians in the room seemed more astonished by the unusually long hairstyle of the geologist interpreting the pictures on the television screen.

When Wilkening returned to LPL, the dean of the College of Liberal Arts was looking for a new director and department head. Tensions between the established astronomers and newcomers at LPL had worsened in her absence, and the dean hoped to find a neutral person to take on the difficult role. Staff and faculty knew Wilkening to be sensible and cool-headed. Although she was still an assistant professor, she became LPL's director in 1981.

NEWCOMERS

Fears that the university couldn't sustain both an astronomy department and the rapidly diversifying Lunar Lab proved to be unfounded. Both institutions thrived, and Wilkening's directorship marked another period of growth. "We had the only department, practically in the world, that was really interdisciplinary and focused on planetary exploration," Wilkening said. "People wanted to come here to work. So it was really easy to hire good people."

One of those was John Lewis, the protégé of Harold Urey who had once visited the lab as an awed graduate student during Kuiper's time. Lewis had found the autocratic director both impressive and unlikable. He turned down Kuiper's job offer and took a position at MIT instead. Sonett later repeated the offer, and Lewis again turned it down.

When Lewis visited LPL the fall of 1980 on sabbatical from MIT, he found an entirely new department. The contention that had marred the previous decade had begun to fade, though it left a mark on the institution that never vanished entirely. Now astronomy blended with physics, geology, chemistry, and engineering, and scientists from each specialty collaborated and criticized in turns. Good-humored arguments spilled out into the halls from open office doors. Graduate students discussed meteorite impacts and mathematical formulas over beers on Friday evenings. Lewis knew and liked Wilkening, who became director during his sabbatical visit: they had shared Urey's mass spectrometer in graduate school at UC San Diego.

Like many of LPL's faculty, Lewis fell in love with planetary science as a child. "Somewhere around the age of three or four I began speculating about what was on the backside of the Moon and stumped my parents, which I found enormously stimulating," he said. The evening before his wedding, Lewis borrowed a friend's television set to watch the impact of Ranger 7. A few years later the young couple bought their first TV to watch the Apollo landings.

Caught up in the excitement of the era, Lewis diligently searched through college catalogs for planetary science classes. There weren't any. Lewis wouldn't find a course on the subject until, years later, he taught one at MIT.

By the time Lewis came to LPL, he knew all of the faculty members and most of the graduate students. He had even taught some of them, including Guy Consolmagno, who had vague plans of becoming a writer before taking Lewis's course in meteorites at MIT. ("The very thought of studying rocks that had been in outer space, that you could actually touch, just absolutely thrilled me," Consolmagno said.)

The convivial atmosphere at LPL—and the clear southwestern skies, so different from bleak Boston—enchanted Lewis. Gazing out the office window at the Catalina Mountains, blue in the spring sky with a dusting of snow, he realized that he didn't want to leave. Wilkening offered him a permanent position the next morning.

Lewis brought to LPL his expertise in the chemistry of the early solar system, focusing on comet and meteorite impacts and extrasolar planets, two topics that would fall under the spotlight in the eighties. After the isolation of MIT, the University of Arizona seemed to offer a "feast of riches" with its cluster of buildings devoted to planetary science, astronomy, and optical sciences—not to mention geology, physics, chemistry, and atmospheric sciences. The prevalence of observatories on nearby peaks meant that Lewis could see almost every working astronomer in the country as they passed through town, once a year at least.

"No matter what I got interested in, in any given week I could find people to talk about it," Lewis said. Scientists at LPL continued to sift through the rich harvest of data returned in the previous decade, and they had engaged in every stage of the Voyager missions, from planning to development to data analysis. The Grand Tour of the solar system was about to begin.

RETURN TO JUPITER

Bradford Smith was already the celebrated head of the Voyager imaging team when Sonett invited him to join LPL in 1974. He had earned a PhD in planetary science from Caltech, where he'd built an impressive record of spacecraft involvement. His frequent appearances in the press made Smith a highly visible and much celebrated representative of the Voyager missions. *People* magazine called him "the nation's tour guide" for the planetary encounters.

Voyagers 1 and 2 sailed past Jupiter in 1979, photographing the marbled patterns of its turbulent atmosphere. The twin spacecraft captured details that the Pioneers couldn't see. As Smith remarked to the press, "the existing atmospheric circulation models have all been shot to hell by Voyager." Strange new satellites appeared, many of them remarkable worlds in their own right. The Voyagers captured the public's imagination like no other mission.

"There was a lot of public support for the space program," Wilkening said. "As long as we were involved and visible in the various space missions, we had the aura of the space program surrounding us. We were much covered and much loved by the local media."

NASA had learned two important lessons from the Pioneers. This time, engineers armored the spacecraft against radiation, and they built in the bandwidth capability to return high-resolution images. The spacecraft's two cameras returned the first close glimpses of the Galilean moons. NASA knew that images would dominate the public's attention and go a long way toward ensuring funding, which threatened to limit the mission's duration.

Engineers had also equipped the spacecraft with almost every scientific instrument imaginable, less flashy than images but more vital for doing good science: photopolarimeters, magnetometers, infrared and ultraviolet spectrometers, and three kinds of particle sensors.

For Tucsonans, the Voyager encounters were a matter of personal pride. LPL, Steward Observatory, and the Optical Sciences Center had set the city on a path to become a lightning rod for good science. Lyle Broadfoot's research group had developed the ultraviolet spectrometers for the two Voyager spacecraft at the Kitt Peak National Observatory, giving Tucsonans a local interest beyond the visible wavelengths. Around the time of the Jupiter encounters, Broadfoot's group became an off-campus facility of the University of Southern California.

Broadfoot had fallen in love with the physics of planetary atmospheres by gazing at the shining ribbons of the aurora in the skies of his homeland, Saskatchewan. He housed his researchers in a small warehouse on Ajo Way in Tucson, setting up clean rooms in the middle of the junkyard district. They stored data on reels of magnetic tape in a double-wide trailer, a wealth of information that today would fit a flash drive. The group, whimsically called "Lyle's Garage," undertook analysis of Voyager data, peering into the emptiness with ultraviolet eyes.

The short wavelengths at this end of the spectrum proved useful for studying hot young stars, stars nearing their death, supernova remnants, galaxies, and other high-temperature objects. The almost-empty space between stars fascinated Jay Holberg, who had charge of the instruments during the cruise stage. Making use of the long journey, he observed the interstellar medium, the continuous material of the universe that fills the gaps between stars.

FIGURE 13. Lyle's Garage, first housed in a warehouse on Ajo Way, became part of LPL in 1983.

COURTESY OF LPL SPACE IMAGERY CENTER

Lyle's Garage would later become an inextricable part of the Lunar Lab's development. In the meantime, the two groups—visible and ultraviolet—looked at Jupiter's satellites with matched astonishment. Broadfoot ruefully recalled dreaming one night, during the hectic days of the Jupiter encounter, that Io was volcanic. He remarked on it to his team, but it hardly seemed worth mentioning to the rest of the Voyager scientists—until the spacecraft's cameras picked up the first sign of smoky plumes rising over the moon's sulfur-coated plains.

Io stunned the watching scientists. The Pioneers had only detected an intriguing red tint to the moon, and few scientists had dared to speculate on active volcanism, wrenched from the moon by the pull of massive Jupiter. The Voyagers observed nine volcanic eruptions, spitting yellow plumes out of an orange-red crust and creating black lakes of liquid sulfur. The bizarre spectral features noted by Fink and Larson in their study of solar system ices turned out to be bright patches of sulfur dioxide frost.

"When they saw Io they were just flabbergasted because it didn't look anything like they expected," said Floyd Herbert, who would join Lyle's Garage for the latter half of the Grand Tour. "It's got all these little volcanoes on it. Brad Smith was the honcho of the imaging team, and he said, 'My God, what kind of satellite is this? I've seen better looking pizzas!'"

Broadfoot's instruments had an important discovery to make. Io's volcanoes spew sulfur and oxygen in plumes so high they escape the moon's weak gravitational field, setting a haze of ionized gas (called plasma) over Jupiter. When charged particles within this circling ring of plasma collide, they emit ultraviolet light. "If we hadn't had an ultraviolet experiment, we wouldn't have a clue and wouldn't know it was there today," Broadfoot said with quiet pride.

STRANGE NEW MOONS

Voyager 1 arrived at Saturn in November 1980, nine months ahead of its sister ship. The ringed planet was the dazzling star of the Grand Tour, butter-gold in photographs with loops of stone and ice dancing attendance around her. The close encounter revealed intricate structures within the rings—lovely spirals, curls, and braids—that hadn't been visible before.

Scientists hoped the two Voyagers would finally explain why the dust orbiting Saturn had shaped into well-defined rings instead of a flat, featureless sheet. Part of the answer came from two new satellites, Prometheus and Pandora, discovered on either side of the narrow F Ring. The moons acted as "shepherds" that herded ring particles into distinctive kinks, braids, and bands, literally keeping them in line.

The Voyagers found no shepherd moons around the biggest and brightest ring, however. With data pouring in at unmanageable volumes, the study of the B Ring's quirks fell to a graduate student at Caltech named Carolyn Porco, who would later join LPL's imaging team. She found an explanation for the mysterious radial features called "spokes" that marred the surface of the B Ring, which astronomers had observed from the ground but dismissed as a fantasy until the Voyagers arrived. Porco discovered the spokes rotated in step with the magnetic field of Saturn. The features quickly formed when the magnetic field swept up charged particles and then faded away again.

The 1980 encounter proved a fortuitous time to make ground-based observations as well because the Earth had lined up with the plane of Saturn's bright rings so they appeared as a thin line. Such opportunities occurred roughly every fourteen years. During the previous "ring plane crossing" in 1966, Andouin Dollfus discovered the tiny satellite Janus.

At LPL, Steve Larson and John Fountain had sifted through Dollfus's data. They became convinced that another satellite, twin to Janus, also orbited close to the rings. They published their findings in 1978 and suffered a barrage of criticism.

The opportunity to prove their claim was just around the corner. In 1980, Saturn again presented its rings edgewise to Earth. Brad Smith had been advocating for the use of charge-coupled device (CCD) cameras at the telescope—brand-new technology at the time. The light-sensitive pixels within a CCD detect the pattern of photons striking the silicon surface, like dimples in a sidewalk collecting rainwater. The cameras were still under development, but Larson, Smith, and Harold Reitsema located a prototype developed for the Hubble Space Telescope. In February 1980, they fitted it to the 61-inch telescope and set their sights on Saturn.

They watched Janus emerge into view, right on schedule. "And then we saw one on the other side come out," Larson recalled in wonder. It was the first observation of co-orbital satellites, two moons sharing one orbit. Locked in its complex dance with Janus, the second satellite had escaped notice until that moment.

Pioneer 11, as it turned out, had passed within a few hundred kilometers of the tiny moon when it flew by Saturn a year earlier. Analysis of Pioneer 11's data revealed that its charged-particle detectors had sensed the moon's wake. It was named Epimetheus, Greek for hindsight.

Newly aware of the moon's existence, NASA instructed Voyager 1 to photograph Epimetheus when the spacecraft arrived at Saturn that November. The twin moons appeared as stony, cratered fragments tangled in the rings. "It was a thrill to be at JPL during the Voyager encounter of Saturn," Larson said. "Not only did they get Epimetheus, they got a series of pictures showing the shadow of the F Ring going across. It was neat to experience a world going from a little point of light to actually seeing it as a chunk with craters"—in Epimetheus's unusual case, a transition that occurred within less than a year.

The moon that most intrigued scientists was Titan because of the tantalizing presence of carbon-hydrogen bonds in its roiling atmosphere. NASA directed Voyager 1 to pass close to Titan to get a better look, although the decision eliminated Pluto as a future target. The maneuver ended in disappointment. Images revealed nothing more than Titan's featureless orange smog, too dense for the spacecraft's instruments to penetrate. Almost at once NASA began planning to send another spacecraft that could delve beneath the obscuring clouds.

Voyager 1's look at Titan did mean a long-awaited confirmation for a theory proposed by Donald M. Hunten, an expert in planetary atmospheres. A Canadian astronomer highly experienced at building instruments to study Earth's aurora, Hunten had been Lyle Broadfoot's mentor in Saskatchewan and joined him at Kitt Peak in the mid-sixties. In Tucson he participated in many NASA committees and played a key role in planning for Pioneer Venus. He sought out a position at LPL during Charles Sonett's directorship.

"From its small beginnings it had grown and flourished and was fostering just the kind of work I wanted to do," Hunten said. LPL offered a chance to refocus his career on planets, still a subject of disinterest among Kitt Peak astronomers. Hunten also wanted to mentor students. He took on graduate students in a number of far-flung research projects, guiding Nick Schneider with his work on the Io plasma torus and Mark Sykes on the dust trails streaming from comets.

Sykes described Hunten as "one of the gods of atmospheric sciences and a brilliant man." Scientists had known that Titan had an atmosphere since Kuiper's detection of methane in 1944. Later observers discovered hydrogen as well, a puzzle because light hydrogen molecules weren't expected to remain on a moon with such weak gravity. Hunten developed an elegant model to explain why hydrogen didn't escape, which is still the standard today. He also suggested that Titan's atmosphere contained nitrogen in the early seventies. The theory, largely ignored at the time, proved prescient when the Voyagers confirmed a dense, nitrogen-rich atmosphere.

The close approach to Titan sent Voyager 1 veering out of the solar system's ecliptic plane. Years later it would plunge through the heliopause, the boundary where the dominion of the Sun fades away and the interstellar medium begins. Voyager 2, on the other hand, could bend around Saturn's gravity at a point that would send it on to Uranus and Neptune with an extra boost of speed. The planets would not be aligned in such a way for another 176 years.

At the launch of the mission, NASA administrators doubted that the money would last long enough to take advantage of the unique positions of the planets. Originally NASA had just enough funding for the flybys of Jupiter and Saturn, but JPL engineers optimistically designed Voyager 2's trajectory to preserve the option of an extended journey. Then photographs of Io's plumes and Titan's haze began appearing in newspapers and on TV screens across America. Everyone could marvel at the mathematical beauty of Saturn's rings, where before only scientists appreciated the exquisite precision of resonating moons and occulting stars.

The public appeal of those images was so intense that Brad Smith became a celebrity. Newspapers around the world wanted his opinion on the unearthly

FIGURE 14. Scientists celebrated with champagne at JPL
during Voyager 1's close approach to Saturn on November 12,
1980. The screen shows a live view of the moon Mimas.

COURTESY OF JOHN SPENCER

photographs. Rumor had it that Frank Zappa asked for his autograph. Suddenly, science was sexy. And funding for the Grand Tour came rolling in.

RULERS AND PEN PLOTTERS

The Voyager encounters, spanning 1980 to 1989, enthralled graduate students in the Department of Planetary Sciences. "We loaded a bunch of people into my family station wagon and drove to California, and hung out at JPL with our jaws hanging down looking at these amazing pictures coming back," Nick Schneider said. "For the later Voyager encounters, I got to go back as a productive graduate student, ruler in hand, measuring the sizes of these geological features that we scarcely understood."

Students took on the often tedious work of analyzing the new data, sometimes armed with no better technology than pencil and paper. Like the Lunar Lab's original students, Schneider and John Spencer counted craters under the tutelage of Robert Strom—this time not on the Earth's Moon but on much more distant ones. Schneider drew by hand the contours of the volcanic plumes erupting over Io's surface, transferring data onto graph paper one pixel at a time.

"It really felt like this was where planetary science was happening," Schneider said. "The field was young; it was broad. We didn't really know what we were going to see as we went out into the solar system." The Grand Tour was their generation's Apollo, the reason they wanted to study the solar system in the first place. Sifting through reams of numerical data, graduate students had the rare opportunity to make fundamental discoveries.

"I remember the first time that happened," Spencer said. "One of the things I did for my dissertation was looking at the temperature measurements of the Galilean satellites made by Voyager during its flyby in 1979—just poking through the data and suddenly realizing, as I plotted points on this Hewlett-Packard pen plotter that we had back then, that the patterns of thermal emissions are totally different between Ganymede and Callisto. Nobody ever knew that before, and here it was appearing before my eyes."

Not all the excitement of graduate school was confined to work, however. When Schneider arrived at Hawthorne House in 1979, two other displaced Wisconsinites already resided there, Gordy Bjoraker and William Merline. Homesick for the Midwest, they decided one October to shop around for bratwurst to share with their friends around a fire. "We jokingly called it the first annual Bratfest," Schneider said. Every year the party grew in size, though the food staples remained constant—beer, brats, cheesecake, and corn on the cob. Now an established tradition among the graduate students, partygoers often number in the hundreds.

The camaraderie of the growing group inspired other events, including a toga party called Bacchanal that took place in the spring. Its major attraction was the bizarre-flavored daiquiris—fish and chips, snow pea and onion—that Merline invented. "[Everyone] got along way better that I would have expected for such a diverse group of people," Merline said. "They were coming from all different kinds of backgrounds—people who really didn't have much in common initially, except for planets. . . . I'm not sure it happens anywhere else."

The students at LPL in the eighties soon gained a reputation as troublemakers. They played endless pranks, filing "magic keys" to fit any door and circulating fake memos on April Fools' Day, which tricked more than one faculty member into a dismayed response. (That tradition still continues, and one such memo, given an obscure and long-winded title typical of scientific papers, can be found archived online at Cornell University Library.)

"One of the things I thought was absolutely delightful was the irreverent group of students that I encountered when I first got to LPL," Humberto Campins said. As a child in Venezuela, Campins dreamed of becoming an astronomer ever since his youthful desire to send a helicopter to Mars. Without a doubt, LPL provided the place to pursue his aspirations surrounded by other graduate students who harbored a childlike love of science.

Mark Skyes arrived in 1981 and focused his studies on infrared observations of asteroids and comets. He was also delighted by the evolving traditions, particularly the impertinent Christmas skits. "We were a pretty close-knit bunch back then," Sykes said. "When I came in, they were also something of an older group. As a consequence we didn't respect authority a whole lot. We gave the poor faculty a bit of a hard time through our tenure."

The teaching staff took it in stride, and the graduate students wreaked havoc at the annual holiday party. In one skit, they solemnly poked fun at Uwe Fink packing the old absorption tube full of explosive mixtures while Godfrey Sill calmly lit his pipe nearby. "We would do things not only to amuse but also to inform," Schneider said. "Sometimes we had to let certain faculty members know that their behavior didn't meet the standards for one reason or another, personal or professional, so we would put that right into the skit."

Hawthorne House was the heart of it all, the core of the students' activities. They slept on the battered couch and gathered there after long days of crunching numbers. On one wall a map of the world bristled with pins that marked the place each student called home, from Tucson to Australia. Over the course of three decades, some seventy students claimed one of Hawthorne House's five rooms as their home, and many others found an informal welcome.

"For a while we were having dinners in the evening on a regular basis," Spencer said. "Or we would just spontaneously say, 'Oh, let's make spaghetti this evening' and we'd all head over there on a Saturday. . . . There was always something going on. Sometimes it was just a couple of people watching TV, drinking a beer in the evening. It was a great place to be social."

The holiday parties and jokes fostered rapport between students and teachers, breaking down the status barriers. On Friday evenings students gathered at the Big A, an old bar on Speedway, with Mike Drake and Brad Smith and whatever other faculty members felt like coming along. Their solidarity formed a unique atmosphere in the department, encouraging open dialogue between students and faculty. Unknowingly, the impudent graduate students had helped create a place where professional disagreements could end up in laughter—as the highlight of a Christmas skit or the target of an April Fools' joke—and ultimately in collaboration.

FIELD TRIPS

The quality of the graduate students impressed Jay Melosh, who was hired by Wilkening in 1982 to bring his much-needed expertise in geophysics to the lab. He specialized in craters and meteorites. When he first arrived at the lab, he focused on solving the puzzle of the SNC (pronounced "snick," it stands for "shergottite, nakhlite, and chassigny") meteorites, which some scientists had suggested came from Mars.

"Gene Shoemaker said in no uncertain terms—I can still remember his voice echoing through the room—that it was absolutely impossible to get a rock off of Mars without it either melting or vaporizing," Melosh remembered. Yet dissolved gases trapped in pockets inside the SNC meteorites were a "dead-ringer" for the Martian atmosphere. Melosh showed how the shockwaves of one enormous impact could have interacted to eject material from Mars, a process called spallation, still accepted as the most likely explanation today.

Such dramatic subjects required equally dramatic scenery, and luckily there were visual aids nearby. Melosh revived a tradition that had fallen by the wayside since Kuiper's time: field trips. A line item in LPL's budget allowed for excursions, but the money hadn't been touched since Kuiper's regular visits to Mexico, Hawaii, and other potential observing sites.

John Spencer attended one of the first field trips, an excursion to Meteor Crater in northern Arizona. "I remember seeing impact melt, real impact melt, there on the crater wall, and being very excited about that," he said. Every few months Melosh chose a location that illustrated a feature found elsewhere in the solar system. "We always have to talk about the planetary connections," he said. "When you look at something on Earth, what are you learning about some process that occurs on a distant planet? Unfortunately we can't take field trips to the Moon and Mars—at least not yet—so we have to do the best we can on Earth."

Melosh required the students to choose a topic and give a talk about some landform along the way. Friends and spouses, tagging along for a weekend of camping, had to make a presentation, too. After nightfall they gathered around a campfire for fireside chats about science, often setting up telescopes and pulling out guitars.

"There are a lot of things you just can't teach by sitting in a classroom," William Bottke said. "I think all of my friends agree that we probably learned more on those field trips than we ever learned in class." After joining the department in 1980, Bottke attended field trips to nearly every major feature within driving distance, from White Sands, New Mexico, to the Blackhawk Landslide in California, to examine the processes that pummel and carve a planet. Students roamed the Pinacate lava fields where Binder, Hartmann, Cruikshank, and Wood used to camp.

"We had this wonderful trip up to the Grand Canyon in February 1981," Spencer said. "About seven or eight of us headed up there in a couple of cars, and we had CB radio going up I-17 toward the Grand Canyon, listening to each other's favorite music very loud on the tape deck. Just that great feeling of camaraderie—I think it was the first time in my life that I felt surrounded by like minds, people who really understood and enjoyed the same kinds of things that I did."

The training and friendships that fostered a strong generation of planetary scientists also made it difficult for them to leave. "It became a very good life as a graduate student, and some would say we got *too* comfortable," Schneider laughed. He recalled designing the Mars Ball, a massive prototype rover balanced on two squishy wheels. Students labored over the project with sewing machines, constructing inflatable tires that could roll over obstacles. Although they proved the design would work, it eventually lost out to smaller, smarter rovers.

The innovative project contributed to the delay in his graduation, and for a while Schneider held the record as the longest-running graduate student at eight-and-a-half years. Bill Merline doubled it. They were not the only ones reluctant to move on. Bottke recalls faculty complaining that the students were so happy they never wanted to leave.

"There was some truth to that," he admitted. "You're not getting paid much money, but you don't have any responsibility except your science. In some ways you have extreme freedom. You're young, you can do what you want, you can work in the science you want. We just had a great time. It probably did take us a year or two longer to get out because we were having too much fun."

FIGURE 15. Graduate students Tom Jones, Bob Marcialis, Mark Sykes, Gene Eplee, and Shelly Pope with the Mars Ball prototype

COURTESY OF JOHN SPENCER

THE FATE OF THE DINOSAURS

Theorists like Jay Melosh and John Lewis who studied impact craters were tackling a topic unpopular with most geologists. Despite the unavoidable evidence of Meteor Crater and the discovery of the SNC meteorites, many scientists believed that celestial impacts had little to do with shaping the biology and geology of present-day Earth.

Still, the idea that an Earth-shattering asteroid could wipe out entire species would not go away entirely. Harold Urey had suggested as early as 1972 that such an event could explain the mass extinction that occurred 65 million years ago, when three-quarters of the Earth's flora and fauna vanished into the fossil record, including, most famously, the dinosaurs.

Few geologists or paleontologists gave Urey's theory any credence. Some favored a series of massive volcanic eruptions as the culprit, but most believed that the extinction event had been a gradual process, occurring over hundreds of thousands of years as evolution weeded out slow, lizard-like species unfit for the competitive world.

The reluctance to consider a catastrophic event stemmed in part from the newness of the planetary science field. Geologists had been trained in a science of slow, visible processes. The drama of a cosmic impact seemed an unnecessarily flashy solution to a down-to-Earth problem. Impact craters, especially as a process that could influence the Earth, had largely been ignored until the Apollo age. Drawing from so many diverse disciplines—a marriage of astronomy and geology—planetary science found little elbow room at the family table of traditional sciences.

The key to proving the impact theory lay within the soil, in the global band of clay that marked the close of the Mesozoic era. In 1980, a team of scientists led by geologist Walter Alvarez and physicist Luis Alvarez reported that the K-Pg boundary separating the Cretaceous and Paleogene periods (formerly called the K-T boundary) was unusually rich in iridium, an element abundant in asteroids and comets but extremely rare in the Earth's crust.

The discovery sparked a worldwide hunt for the impact crater that sealed the fate of the dinosaurs. When Urey first postulated a calamitous space collision, he suggested that geologists should look for tektites, rounded droplets of dark glass with distinct differences from the obsidian found in volcanoes. In Urey's time many scientists (including Kuiper) believed that tektites came from the Moon, a hypothesis that the Surveyor landers laid to rest, much to Urey's satisfaction. A new consensus emerged that tektites form when an impact sends molten drops skyward, which harden into aerodynamic shapes as they spin through the atmosphere.

Tektites were one clue among many in a planetary-sized jigsaw puzzle. Scientists had thought they found minuscule tektites in the K-Pg layer as early as 1981, as well as confirmed dozens of iridium-rich sites around the world, but the evidence remained shaky. It was the perfect time for a young Canadian geologist named Alan Hildebrand to join LPL's ranks, intending to focus his graduate studies on asteroids. Instead, he decided to solve the mystery.

In 1984, the same year that Hildebrand arrived in Tucson, a USGS geologist announced the most important evidence for the impact hypothesis yet: shocked quartz grains in the K-Pg layer in Montana, their internal structure deformed by intense pressure into a crosshatched pattern characteristic of meteorite impacts. Hildebrand, a field geologist by training, traveled around the world to bring back samples from known K-Pg boundary sites, spreading plastic bags of soil across a table in his laboratory. He hunted particularly for impact wave samples—sandy debris tossed up by the tsunamis that would have occurred if an asteroid hit the ocean—and found them at the Brazos River in Texas and along the Gulf of Mexico.

The Earth had changed greatly since the Cretaceous period. Subducting continents could have easily obliterated the evidence. Odds favored an ocean impact—after all, water covers two-thirds of the planet—but the chemical signature of the samples confused scientists. Some clues, such as shocked quartz, pointed to a terrestrial impact, while others indicated an ocean target.

Then Haiti caught Hildebrand's eye. A geologist there had reported volcanic deposits at the K-Pg boundary layer. After visiting the site with fellow student David Kring, Hildebrand became convinced the deposit was actually impact ejecta. It lay beneath the global "fireball layer" tossed up by the impact, only about three millimeters thick and chock-full of telltale siderophile (iron-loving) elements like iridium. The fireball layer had been dispersed globally when the asteroid or comet vaporized on impact, but the heavier ejecta layer—rock and earth flung outward by the collision—had largely settled between North and South America.

The ejecta layer in Haiti lay half a meter deep, considerably thicker than ejecta found at any other known K-Pg site, complete with tektites and grains of shocked quartz up to a centimeter in size. The crater had to be nearby.

In an odd twist, it *was* nearby—and had already been discovered. In the fifties an oil geologist working for Petróleos Mexicanos (Pemex) on the Yucatán Peninsula noted a circular anomaly at a drill site and wondered if it was an impact crater. The information was proprietary and couldn't be made public. Then, in the late seventies, an American consultant to the company named Glen Penfield identified magnetic anomalies at the same site. He convinced the company to let him (along with Mexican geophysicist Antonio Camargo) report the finding at the annual meeting of the Society of Exploration Geophysicists in Los Angeles.

Penfield and Camargo understood the discovery's importance. The data suggested a ring of once-molten rock 180 kilometers across, with a dense uplift in the center: exactly the size as the proposed K-Pg boundary crater. Its age was correct as well, at 65 million years. The feature lay half on land, half in water. If an asteroid or comet had struck in the shallow water of a continental platform, that could explain the mixed signals in samples around the world.

The significance was not lost on the popular press. Carlos Byars of the *Houston Chronicle* ran a front-page story linking the Yucatán crater to the dinosaurs' demise. Strangely, however, the scientific community remained silent. Penfield contacted experts at NASA's Johnson Space Center and Walter Alvarez investigated his claim, but the difficulty of obtaining samples from the structure, buried a thousand feet below limestone, discouraged interest.

The Yucatán anomaly had nearly been forgotten by the time Hildebrand examined the ejecta layer in Haiti and calculated that the crater lay within one thousand kilometers of Haiti's position during the Late Cretaceous period. Byars, the Houston reporter, first suggested to Hildebrand that he call the Johnson Space Center and see if anyone remembered the name of the Pemex geologist who had made a major find in the Yucatán. Hildebrand took the advice, and his search at last led him to the listing for Glen Penfield in the Houston phonebook.

"At the time it was just one more suggestion for the impact crater," Hildebrand said. "When I first talked to Glen he said something like, 'We're not sure if it's the crater. I haven't done anything about it for years.'"

The Yucatán anomaly lay just outside the circle Hildebrand had drawn around Haiti. Hildebrand needed samples to confirm that a crater lay beneath, but Pemex had lost many of its drill cores in a warehouse fire. With Penfield's help, Hildebrand located a few samples drilled from outside the supposed crater's walls, which showed evidence of the ejecta blanket.

Meanwhile, journals rejected the first three papers that Hildebrand wrote about the discovery. Nobody in the conventional scientific world wanted to listen to a graduate student and an oil geologist. Hildebrand finally located samples from inside the crater, courtesy of Camargo. "It turned out he'd had a couple of samples sitting on his desk for a decade that had been drilled from wells inside the crater from the 1950s and 1970s," Hildebrand said. "They had evidence of shocked metamorphism in them, too. That was the confirmation."

Hildebrand enlisted the help of Kring, Penfield, and Camargo to write the paper, together with his thesis advisor Bill Boynton, Canadian geophysicist Mark Pilkington, and Harvard geochemist Stein Jacobsen. It appeared in the September 1991 issue of *Geology*. He named the crater after a nearby town, Chicxulub, a Mayan word meaning "tail of the devil."

By the time Hildebrand defended his thesis in December 1991, there was little doubt that he had identified one of the largest impact craters on Earth.

Within a few short years, long-divided camps of geologists, paleontologists, and planetary scientists agreed that Chicxulub marked the long-sought site of the K-Pg catastrophe.

The leap from impact to extinction still had to be made, and other LPL scientists took up the challenge. As early as 1982, John Lewis had published work suggesting that the K-Pg impact could have produced acid rain, dissolving the calcium-based shells of sea creatures. Later calculations by Jay Melosh suggested that in the first few hours after impact a storm of glowing fireballs bombarded the Earth, leading to a massive die-off of phytoplankton in the boiling surface waters. In both cases, the effects would have rippled up the food chain.

Hildebrand summed up the theories neatly when he wrote that the impact "turned the Earth's surface into a living hell, a dark, burning, sulphurous world where all the rules governing survival of the fittest changed in minutes. The dinosaurs never had a chance."

WATCHING THE SKIES

Another critical event occurred in the eighties that changed the way scientists thought about impact craters. Ever since William Hartmann and Don Davis had published their giant impact theory in 1975, the question of the Moon's origin had remained open. No one yet agreed on whether the Moon had been captured by Earth's gravity, spun off in molten form at the Earth's birth, or smashed into existence by an enormous collision. In October 1984, Hartmann helped convene a conference to hash out the issue in Kona, Hawaii. To everyone's astonishment, speaker after speaker stood up to give evidence in support of the giant impact theory.

The Alvarez discovery and the resulting hunt for Chicxulub, in part, had smoothed the way for this remarkable consensus on the importance of cosmic collisions in shaping the solar system. It was clear, at last, that catastrophic impacts could dramatically reshape Earth's geology and biology. Even before Hildebrand had settled the fate of the dinosaurs, astronomers at LPL had set up a program to make sure humankind didn't go the same way.

In 1980, Tom Gehrels wrote a draft proposal for the Spacewatch program, which would search the solar system for asteroids and comets that might present a hazard to Earth. The proposal showed remarkable foresight, given the developments to come in that decade that would revolutionize ideas about catastrophic impacts. Gehrels showed it to Robert McMillan for advice. McMillan wrote back a stinging criticism, identifying the project's major problems.

"To my astonishment, he made me the deputy investigator of Spacewatch," McMillan said. Over the next decade, the two scientists worked to develop the electronic techniques that would allow for systematic detection of near-Earth objects (NEOs). To develop scanning techniques with CCD cameras, Gehrels adopted the venerable Steward Observatory telescope that once stood on campus before light pollution from the sprawling city forced its relocation to Kitt Peak. In 1989, the Spacewatch program detected the first NEO found with a CCD camera.

Spacewatch observers spend long nights on the peak, starting at 6 p.m. and staying vigilant until dawn. The computer scans slivers of the night sky three times in succession to look for fast-moving objects. The observers also keep watch in real time, looking for telltale streaks moving against the predictable background of stars. They then need several more hours to sort through thousands of snapshots called up by the computer program, each one a possible find.

"I'm quite proud of the accomplishments of Spacewatch," McMillan said. "We have a certain niche in the world that nobody else is doing." The program reports roughly two dozen new NEOs each year, a significant contribution to safeguarding the planet against hazards. Along the way, Spacewatch discovers thousands of new asteroids, gathering data to inform studies of the solar system's evolution and to suggest targets for future spacecraft missions.

Gehrels passed the program's directorship to McMillan in 1997, and it became his life's work. By then, Spacewatch was no longer alone in its guardianship of Earth: similar programs had been initiated in Flagstaff, New Mexico, and Hawaii, and amateur astronomers continued to help in the hunt. McMillian steered Spacewatch toward following up objects after they become too faint for survey telescopes to track, an unusual task for such high-quality technology. "We actually have the largest telescope in the world that is dedicated full-time to searching for asteroids and following them up," McMillan said.

HALLEY'S COMET RETURNS

Although LPL had relinquished management of the Catalina Observatories, ground-based research remained a crucial facet of the lab's activities. Telescopic techniques made a quiet counterpart to the flashy spacecraft programs, filling the gaps between fluctuating funding and providing vital data that scientists couldn't collect any other way.

The spectacular Comet Bennett graced the night skies of 1969–70, displaying remarkable structures that Steve Larson had only seen in drawings. "I thought they were figments of someone's imagination," Larson said. "There were spirals and all kinds of fantastic features, and the comets I had seen up to that point were just fuzzy things. But Comet Bennett came along, and by God, there were these spirals. That just absolutely blew me away."

Larson's first love had always been comets. As a child he had marveled at the dazzling light of Comet Mrkos, bright enough to see with the naked eye. "We lived out in the desert, so I had a good view of this thing," Larson recalled. "That was just about the time the Space Age started, and there were satellites—I used to go out and look at those."

Larson rekindled his love of comets at LPL, first characterizing the structures in Comet Bennett and then joining a subdivision of the International Halley Watch. He never got a PhD. In the highly stratified society of academia, he was told at one point that he didn't have a future in planetary science. He ignored the warning. With Wilkening's encouragement, Larson soon began writing proposals for his own funding. As a result of his work on Comet Bennett, he was invited to join the Vega Program, two Soviet spacecraft that flew by Halley's Comet in 1986.

Although the United States had canceled its comet mission, a flotilla of spacecraft from other agencies—Soviet, Japanese, and European—would intercept Halley's Comet, which was making its first comeback since 1910. Meanwhile, astronomers around the world made arrangements to observe the comet from the ground. Larson enlisted the help of Canadian astronomer David Levy and obtained a bulky CCD camera for the occasion. Levy had also hunted comets since childhood, but he chose to pursue his education in English, finding the poetry of the night sky in Shakespeare. Now a renowned author and astronomer, he has discovered twenty-two comets.

Larson and Levy mounted their CCD camera to the 61-inch telescope in the Catalina Mountains, cooling the sensitive instrument with liquid nitrogen. The imaging technology was so new that Levy ran to a friend's office to ask in embarrassment, "What's a pixel?" In March 1986, Larson shipped the portable CCD camera to South Africa, which had the best views of the comet. He flew from South Africa to Moscow to watch Vega 1's anticipated rendezvous with

Halley's Comet. In the space of three hours, Vega 1 returned some five hundred images of the comet's nucleus and glowing coma, a fiery cloud of ice and dust.

CCDs could record starlight much quicker than the photographic techniques they replaced, as well as keep a digital record. The old style of observing, it seemed, was vanishing. Computerized systems crept into the telescope domes, and astronomers no longer needed to spend their nights on the observing platform, staring up into the spangled night. Increasingly, planetary sci·entists turned telescopes to the sky to lend support to spacecraft missions, no longer expecting to make profound discoveries from the ground.

The trend toward reliance on spacecraft, however, did not continue unchecked. The Voyager missions wound down in the late eighties, and funding for planetary research began to trickle away. The Apollo fervor was gone. America had conquered the Moon, and Mars's great mystery had stripped away under the unforgiving cameras. Once Voyager completed its final circuit, all the planets in the solar system barring Pluto—not yet demoted from that status—would have received at least one robotic visitor.

With spacecraft work trickling away, some planetary scientists returned to the telescope lens with a weighty question in mind: did planets exist around other stars? Had the field of planetary science found its limit at the edge of the solar system—or did something wait to be discovered beyond the vast emptiness that surrounded our Sun?

PLANETS BEYOND PLUTO

The dream of finding exoplanets (planets orbiting other stars) dominated the last years of Krzysztof Serkowski's life. A Polish astronomer, Serkowksi had fled his home country during the Cold War. At Lowell Observatory in Flagstaff he became known for his work on the polarization of stars, and Tom Gehrels invited Serkowski to join his project at LPL in 1970.

Serkowksi designed an exquisitely sensitive instrument to detect the telltale variations in the velocity of stars that indicate an orbiting planet. The method relied on the fact that planets exert a small but discernible force on their sun's orbit. As a result of this wobble, the star spins sometimes closer to Earth and sometimes farther away. An observer on Earth saw the star's spectral lines shift toward red when the star receded and toward blue when the star approached.

By measuring this movement, an astronomer could estimate the mass of the orbiting planet.

Serkowski believed that the search for life beyond Earth hinged on the discovery of exoplanets, only a theory at the time. Carl Sagan, the renowned scientist and writer who had studied under Kuiper at Yerkes, had popularized the idea of these distant, unknown worlds, but many astronomers remained skeptical of their existence.

Serkowski dedicated the last days of his career to the search. He speculated that exoplanets might bear entire civilizations that knew the secret of surviving the advent of technology, which Serkowski had seen applied for heartbreaking destruction in his home country. "We should not ignore the possibility that searching for extrasolar planets may be more important to the survival of mankind than any other human activity," he wrote in 1977.

By then, Serkowski already knew he was dying of amyotrophic lateral sclerosis. He continued to work from his bed at home, boundlessly energetic. When Serkowski passed away in 1981, Robert McMillan inherited his Radial Velocity Project. He remembered Serkowski as an astronomer of the old school "who personally went to the telescope and spent all night with it . . . no control room, just standing on a platform all night running the equipment."

Determined to honor Serkowski's legacy, McMillan wrote proposals for funding that candidly explained the problems the program had encountered. He received grants from the National Science Foundation and NASA. McMillan joined forces with Peter Smith, who was looking for a project to keep busy with during the dry time in the eighties.

Smith and McMillan, aided by graduate students, redesigned Serkowski's instrument and built it on the loading dock behind the Kuiper Building. It could detect a star's wobble down to ten meters per second, sensitive enough to detect big, Jupiter-like planets revolving around distant suns. They choose Arcturus for their first test with a modest 14-inch Celestron telescope, and were taken aback to find enormous jumps in the star's velocity.

"It turns out that Arcturus is not a star that just sits there; it's pulsing," Smith said. "Nobody knew that; this was the most sensitive instrument ever built."

Convinced at last the instrument worked correctly, they took it to Kitt Peak. Two weeks every month, on alternating shifts, they observed during the bright time when the Moon dominated the sky. They shared the telescope with Gehrels, who merited the dark time because he was looking for faint asteroids with Spacewatch.

"You'd be up there at this telescope all alone," Smith said. "There was nobody but you in this big dome building. We only observed in the winter, so it was fourteen-hour nights. You'd set everything up and start your observing. At about four in the morning, you'd battle to stay awake."

Every hour Smith shifted the telescope's position, his eyes on the screen that displayed the image from the entrance slit of the instrument. Pushing the telescope drive buttons, he fought to keep the targeted star in the center of the field of view, "like the dullest video game you could imagine," he said. Sometimes near the end of a long shift, dizzy with fatigue, he stopped to walk outside. The big telescope dome jutted out above the dark juniper trees, and for miles around the peak he could see nothing but the emptiness of a midnight desert.

"There's a little place at the top of the telescope where, if you climb all these ladders, you can stick your head out and see the sky," Smith said.

And you could actually stay out there like this, in a little crow's nest. I remember one time walking up all those stairs, trying to stay awake, and Halley's Comet was visible, rising in the east. I was hanging on, half-conscious because I was so tired, and it felt like I was on a ship. I could feel myself moving. I think the wind was blowing or something. It was amazing because the stars were so bright and beautiful. That's the joy of astronomy, I think. You see those stars and of course wonder about them and you've got your instrument down there and you try to see if there's a planet on that one.

Unfortunately, they weren't lucky with their choice of targets. Serkowski's instrument worked perfectly, but it required vast amounts of light, a limitation McMillan called "photon starvation." They had enough telescope time to look at twenty stars.

"Now it turns out that for about every hundred stars, one has a planet," Smith said, an estimate on par with Kuiper's prediction in the forties. "We only had twenty. We had about a fifth of them." The existence of worlds beyond our solar system wasn't proven until 1992, when two radio astronomers found a planet orbiting a pulsar. Three years later, observers in Switzerland discovered the first Jupiter-like planet around an ordinary star. Six exoplanets were found the following year, all by measuring the subtle wobble of the stars' velocities.

Serkowski's name is still remembered in the field of polarimetry, attached to a law that describes the behavior of starlight as it travels through space. Few recall his commitment to the search for other worlds. Today nearly two thousand exoplanets have been confirmed, with thousands more under investigation—a plethora of possibilities that would have delighted Serkowski, particularly because the majority of early discoveries came from his radial velocity technique.

THE GRAND TOUR

In 1983, as Voyager 2 embarked on the long journey to Uranus and Neptune, Laurel Wilkening resigned from the position of department head and director of LPL. Her skills had not been lost on university administrators, and she rapidly ascended into prominent positions. Shortly after she became director, President Schaefer appointed her to the newly created position of dean of sciences. In 1985 she became the vice president of research—the first woman to hold that position—and the dean of the graduate school. The same year, President Reagan appointed her as the vice chairman of the National Commission on Space.

Now Wilkening was ready to leave her research career at LPL to pursue administrative positions. She became provost at the University of Washington— again, the first woman to hold that position—and then a strong-willed chancellor at the University of California, Irvine. Her good humor had eased LPL through the acrimony developing from the split in disciplines, and she won herself lasting friends with her championship for women in the sciences. For years afterward, faculty reflected on her departure with both admiration and regret.

In their usual irreverent fashion, graduate students commemorated the transition by plastering the halls with campaign posters for various faculty members. LPL's directorship passed to Eugene Levy, an astrophysicist who had arrived at the lab in the mid-seventies with interests in cosmic rays and magnetic fields. When the university offered Levy the position of acting director, he refused. He was unwilling to leave LPL in limbo without permanent leadership, so he suggested that Wilkening remain while the university conducted an accelerated search. Wilkening agreed, and Levy accepted his appointment as director and head in 1983.

"I had programmatic aspirations in some large sense," Levy said, "one of which was to carve out a major place in LPL in spacecraft experimental work. I think that's been an extraordinary success."

LPL adopted Lyle's Garage the same year, an arrangement that gave LPL an excellent group of experimenters while giving Broadfoot's group space and solid funding. Bursting with new personnel, the lab took over the ninth floor of the Gould-Simpson Building with its gorgeous views of the mountains, calling it "LPL West." Despite the impressive new premises, Lyle's Garage never shed its humble nickname. Floyd Herbert joined the group just before the second set of planetary encounters with Uranus and Neptune.

"Those two planetary encounters were the high point of my career," Herbert said, "because they were such a scientifically revolutionary time." Unlike Jupiter and Saturn, scientists knew virtually nothing about this second set of planets. Uranus gleamed as a mere greenish speck in the lenses of the Catalina telescopes. It had five known moons before the Voyager 2 encounter, and only in the last decade had observers discovered a system of narrow rings.

Voyager 2 made its close approach to Uranus in January 1986. Carolyn Porco, now a member of Brad Smith's imaging team, modeled how Uranus's moons (which now numbered fifteen) exert force on the thin, soot-black rings, confining them into particular orbits. Broadfoot's spectrometers detected the ultraviolet dayglow of Uranus's sunlight pole. Floyd Herbert and Bill Sandel worked to pinpoint the elusive auroras created by charged particles skipping off the planet's atmosphere, which were sent crazily awry by the corkscrew magnetic field.

Miranda, the satellite discovered by Kuiper in the forties, surprised scientists with its odd terraces and canyons, features impossible to see from Earth. Ewen Whitaker and Richard Greenberg had measured Miranda's eccentric orbit years earlier by analyzing old photographic plates stored in LPL's vault. Voyager 2 now proved their calculations correct.

Thus ground-based work remained important even in the midst of Voyager 2's encounters with the least-known planets in the solar system. Neptune was Voyager 2's next and final stop. From Earth, Neptune is never visible to the unaided eye. Astronomers had identified only two of its satellites by the time Voyager 2 launched in 1977.

Triton, the largest satellite, had been known since the nineteenth century: it was found not long after Neptune itself. Kuiper made the next finding, Nereid, in 1948. A third discovery, Larissa, happened at LPL during Voyager 2's cruise to the outer planets. William Hubbard, Harold Reitsema, Larry Lebofsky, and David Tholen made plans to study Neptune's rings in 1981 by watching the planet pass in front of distant stars, a process called occultation. They set up the experiment at Kuiper's 61-inch telescope on Mt. Bigelow and the 40-inch on Mt. Lemmon. "The planet Neptune has been one of my lucky planets," Hubbard said. "We saw a drop-out in our signal on both telescopes, and when we analyzed it, we con-cluded that we discovered a previously undetected satellite of Neptune."

When Voyager 2 arrived at Neptune on August 25, 1989, it observed this third satellite, Larissa, and discovered five other moons not observable from the ground. The spacecraft confirmed that Larissa was slowly spiraling inward, fated to eventually impact Neptune or break up in a ring of debris. Hubbard proudly hung an image of Larissa (not yet given that name) in his LPL office, noting that the discovery of the first Neptunian moon after Nereid seemed a fitting tribute to the telescope Kuiper had cherished so much. "It's kind of like deep sea fishing," he said. "You have this long line out into space and you don't know what you're going to reel in."

Scientists gathered an astonishing amount of new data during the encoun-ters with Uranus and Neptune. But the public, anticipating more dramatic images of broiling storms and rocky rings, largely found these last two planets disappointing. Smooth and featureless, Uranus revealed little about its aqua-marine surface. Plain blue Neptune wasn't much better.

The moons, however, were another matter. Sandel, a longtime member of Lyle's Garage, recalled Triton as one of the high points of the long mission. Broadfoot's group performed an occultation experiment, tuning their ultra-violet spectrometer to the changing spectrum as Triton passed in front of the Sun. "I realized at that moment I was the only person in the world that knew what the major constituent of the atmosphere of Triton was. Wow," Sandel said. "But then I blabbed it, so that only lasted for about five minutes."

The Triton encounter caught the public's imagination in a way that Uranus and Neptune had failed to do. In the pale light of false dawn, news crews filled the streets in front of JPL. Scientists poured over the cantaloupe-patterned ter-rain, glassy lakes of frozen nitrogen, and active plumes of nitrogen gas. Over-night, Triton had transformed from a pinpoint of light to a swirling, active world, and newly discovered satellites emerged alongside of Neptune's pale bulk. This was Voyager 2's last stop. It would follow its sister ship to the bound-ary of the heliosphere, beginning the long plunge out of the solar system.

"It was like walking out of the darkness into a clearly illuminated space, because we knew almost nothing, and then in the space of two or three days suddenly we knew an enormous amount about these planets," Herbert said. "It was really an incredible experience. It was like you've been wandering around in a darkened forest and suddenly you come out into Yosemite Valley in bright sunlight."

PART FIVE

RETURNING

PLUTO'S RECLASSIFICATION as a dwarf planet in 2006 retroactively awarded Voyager 2 the honor of completing the first assessment of every planet in the solar system. Streams of data, wondrous in their own right, replaced fantasy. It was clear that Earth alone in the solar system supported the intelligent life that Percival Lowell once imagined on Mars. Voyager 2's encounter with Uranus in 1989 marked the end of a golden era. NASA had plans underway for new missions—Galileo to Jupiter, Magellan to Venus, and the ambitious Cassini-Huygens to Titan—but launch windows slipped by unused as funding snarls diminished cherished plans.

The late eighties were a dry time in space exploration. Eugene Levy envisioned a continuing role for LPL in spacecraft work, not just for scientists who analyzed data but also for engineers who designed and built instruments. Future exploration would require probes of increasing complexity. It was no longer sufficient to merely fly around a planet, capturing those astonishing first sights. Return trips would not be so easy.

When the Challenger Space Shuttle disintegrated just a minute into its flight on January 28, 1986, killing its seven passengers—including a high school teacher—the American space program seemed to have come to a catastrophic end. Graduate students, wandering the hushed halls of the Kuiper Building with television sets blaring out from every office, wondered if they would find any jobs waiting for them in their chosen field.

Like Kuiper at the close of the Apollo era, Levy turned his attention toward education. The Department of Planetary Sciences remained primarily a PhD program to ensure that prospective candidates would have foundations in a broad base of disciplines. It did not offer an undergraduate degree. The university wanted research departments to take a stronger role in undergraduate education, which suffered from large class sizes and a shortage of teachers.

"I was particularly interested in fostering the department's role in the teaching program in the university," Levy said. "It took some time, I think, to engender the sense of commitment to the undergraduate program in the department and the laboratory."

Several faculty members, however, took up the challenge with gusto. Harold Larson was casting around for something new to occupy his time when the KAO program ended in the early nineties. When Levy asked him to teach an undergraduate class, he discovered that the overcrowded classroom conditions left a lot to be desired. "I vowed at the end of the semester that I was never going to teach a class that way again," Larson said, "just lecturing with virtually no way to enhance the learning environment."

That experience sparked the beginning of the Teaching Teams Program, finalized in 1998. With federal funding, Larson and a group of professors from different departments revamped the style of undergraduate classrooms, focusing on hands-on activities and individual interaction. The program, Larson said, has never been turned down for a grant.

Larry Lebofsky also had been dismayed by his first experience teaching undergraduates. With a grant from the National Science Foundation, he established Project ARTIST (Astronomy-Related Teacher In-Service Training) to provide curriculum materials and training to teachers. "You're not going to get good students unless you provide better training for the kids at a younger level," he said. "When I'm looking in my classes I always grab ahold, so to speak, of my future teachers." He remembers the heydays of Project ARTIST as a highlight of his career.

The crown jewel of Levy's new education focus was a grant from the National Space Grant College and Fellowship Program, which Congress established in 1988 to muster the resources of space-centered institutions for the benefit of American students.

Only one institution per state would receive funding in the initial call for proposals. Rather than competing with fellow universities, Levy arranged a collaborative effort between Arizona State University, Northern Arizona University, and the University of Arizona. He spearheaded the proposal and won funds for the Arizona Space Grant Consortium in the program's first competitive round, along with twenty other independent consortia around the country.

Today, Space Grant programs exist in every U.S. state, Washington, DC, and Puerto Rico. The Arizona consortium has grown from the original three universities to thirty-two members and partners. The program sponsors a broad range of educational programs, but its flagship is the statewide Undergraduate Research Internship Program, which partners undergraduate students with mentors across the state to integrate them into Arizona's research enterprises.

"There are too many success stories to count," said Susan Brew, who manages the program. "You name a mission or a significant grant that has a space sciences or an earth sciences component, and Space Grant students have contributed. Since it began, Arizona's has always been a top-ranked Space Grant program in the nation."

When Chris Impey, assistant professor of astronomy at Steward Observatory, called Brew in 1991 to ask if she would interview for the newly created position of Space Grant program manager, she responded, "I think you must have the wrong person." She had worked in archeology and applied anthropology at the Arizona State Museum before the closure of the contract archeology office left her looking for a job. She agreed to the interview, however.

Levy and Impey explained their vision for Space Grant and the need for an enthusiastic manager to build the brand-new program and handle the day-to-day work, freeing Levy to focus on his other responsibilities. "We talked for hours," Brew recalled. "It was a wonderful afternoon. . . . The idea was not to spend a great deal of Space Grant money on building a bureaucracy, but rather on providing exceptional opportunities to students."

Brew notes that the program selects scholars not necessarily on the basis of grades, but on their motivation for learning, "students that we think could find such an opportunity transformational." It seeks nontraditional and minority students who may have never had a chance to discover a love of science and engineering. Space Grant alumni would soon be driving rovers on Mars, investigating near-Earth asteroids, and running million-dollar NASA missions.

The new focus on education had to find a home in the old building. NASA had invested in the Space Sciences Building with research in mind and had forbidden classrooms in the original design. That was remedied in 1993 when the university completed an annex, closing in the old instrument-building bay and taking over the back parking lot. The following year, after a decade of serving as head and director, Levy stepped down to accept a position as dean of the College of Science, and Mike Drake took the reins.

NAMETAGS

The Kuiper Building, which had amply provided for the early lab, was hopelessly cramped before the construction of the annex. Graduate students, about two dozen in all, crowded into a single room behind the business office. They called it "the ghetto" and displayed their nametags on the outside of each tiny cubicle. Uwe Fink still experimented with the building's old absorption tube, which ran right through the offices.

"Typically there were two of us to a cubicle," Paul Geissler said, "and the doors were decorated with all kinds of outrageous stuff, clippings and preprints and cartoons. That forged some solidarity among the graduate students because you're literally on top of each together."

Since the department's founding in 1973, several generations of students had passed through LPL's halls, many of them adopting Hawthorne House as home. Their years at graduate school were not necessarily easy. They worked long hours for little pay, and though stipends rapidly improved over time, some students recalled living on food stamps and boxes of macaroni. Yet an inexpressible pride in the field they had joined buoyed their spirits.

Nick Schneider, earning his PhD a decade before Geissler arrived, remembered his delight at joining the ranks of scientists graduating from LPL. As each student received freshly inked diplomas in planetary science, they peeled their nametags from the cubicle walls and stuck them to a bulletin board. The tradition began with Wayne Slattery, the department's first official graduate, and continued for more than a decade.

"We knew that we were building the field of planetary science," Schneider said. "That list now is probably ten times longer than when I was there. It was the great ritual—pulling my nametag off the door and putting it on the list of PhDs from this department."

Incoming students continued to shape the direction of LPL's many-faceted research programs, pursuing topics ranging from the windblown dunes of Mars to Io's volcanoes to the sheen of methane ice on Pluto. John Keller applied to the program in 2002 after teaching high school and middle school for eight years. The faculty convened a special meeting to review his application. "I was kind of like an experiment," Keller said. "They had never officially admitted somebody doing education research." He soon found a place at LPL on spacecraft missions that increasingly needed expertise in education and public outreach.

Graduate students continued to uphold the department's traditions, chatting around campfires on field trips, cooking bratwurst each October, and circulating mischievous memos on April Fools' Day. They continued, too, the legacy of playing pranks: one year during Drake's directorship, students hid Easter eggs in his office, one of which was never found.

Memorabilia still decorates the hall on the north side of the Kuiper Building, where the graduate students now reside. The added expansion in 1993 meant that bigger and better offices became available, and the ghetto was destroyed along with the nearly three-decades-old absorption tube that ran the length of the building. The bulletin board that carried the nametags of every graduate student since the department's formation also disappeared.

A CLOSER LOOK

After the Challenger disaster, NASA cancelled space shuttle launches for the next three years to take a hard look at safety protocols. That meant delaying Magellan and Galileo and subjecting both spacecraft missions to longer transit times, with all the worry that entailed. Magellan's launch to Venus in May 1989 marked America's first blast-off into the solar system since the Pioneer Venus orbiter and multiprobe more than a decade earlier.

Magellan settled into orbit around Venus in August 1990. The lightweight and economical spacecraft, scrapped from spare parts, carried only one experiment: a radar sensor. Radar penetrated the thick cloud layers and bounced back from the ground, allowing Magellan to map the hidden surface. Victor Baker, an LPL geomorphologist who served on the Radar Investigation Group, said the images were of "immense scientific value." They revealed how past volcanic activity had smoothed and reshaped the planet's young features.

Despite the success of the four-year mission, the space program's comeback was slowgoing. Funding for Magellan ebbed away, and NASA eventually canceled the data analysis program. Theorists, teachers, and traditional astronomers at LPL had plenty to do, but instrument-builders and experimentation groups, which LPL's directors had painstakingly recruited, watched federal funding evaporate with growing apprehension.

Bill Boynton had a string of bad luck during this time. Boynton had arrived at LPL in 1977 with the intention of setting up a laboratory for studying meteorites. He fell into the thorny world of spacecraft instrumentation when NASA selected him to lead the gamma ray spectrometer team for Mars Observer. Gamma rays, the shortest wavelength on the electromagnetic spectrum, emit from excited nuclei in the Martian surface, which is continually bombarded by cosmic rays. By measuring this gamma radiation from orbit, the instrument could map common elements such as carbon, calcium, iron, oxygen, and sulfur. Paired with a neutron detector, the instrument could also identify hydrogen— the key in NASA's search for water.

That question remained unanswered. Approaching Mars on August 21, 1993, Mars Observer inexplicably fell silent three days before its scheduled arrival. Reviewing the failure, NASA officials concluded that a small amount of propellant had leaked and exploded when the engines were pressurized in preparation for entering orbit.

Boynton's profound disappointment in this loss was underscored by the fate of another mission in the planning stages, the Comet Rendezvous Asteroid Flyby, a spacecraft meant to sail alongside of Comet Kopff on a three-year journey. Boynton and his research team had intended to supply the penetrator that would carry a spectrometer into the comet's icy surface. Luck ran against him again. NASA cancelled the flyby, pulling its funding to make room for the ambitious Cassini-Huygens mission to Saturn and Titan.

For scientists and engineers not directly involved with the international Cassini-Huygens project, the late eighties and early nineties proved difficult. Failed spacecraft, cancelled missions, and rejected proposals added up to years of work without a single publication. Struggling to emerge from the cloud of the Challenger disaster, the space program floundered.

The Galileo mission, intended for Jupiter, was severely delayed. Its experimenters joked that they had signed onto the project as fresh young faces and would be close to retiring before they got the data back. When Galileo finally lifted off in October 1989, new safety protocols meant the spacecraft had to launch on a lower performance rocket than originally intended. Galileo needed several gravity assists to make it to Jupiter, lengthening its journey to six years.

At LPL, Richard Greenberg waited to receive the data with a team of dedicated students on hand. The group included graduates and postdoctoral associates such as Paul Geissler, Dan Durda, Greg Hoppa, Dave O'Brien, and Karen Meyers. Exceptional undergraduates found a place on the team as well, including Space Grant interns Terry Hurford and Alyssa Sarid. Randy Tufts was the only classical geologist on the team. He had arrived late at a career in space exploration, already famous as codiscover of Kartchner Caverns, a living limestone cave south of Tucson that Tufts kept secret for more than a decade until he could ensure its protection as a state park.

For these students, Galileo provided much-needed opportunities to break into careers in space exploration. The gravity assists meant that the spacecraft would make two close flybys of Earth, gathering data that many established scientists had no interest in studying. The task fell to Geissler, who had studied seismology in Australia before opening a magazine on a train, reading an article about Uranus, and abruptly changing his field. Alone in the command center in Pasadena one day at lunchtime, watching the familiar images of Earth flicker across the computer screen, Geissler looked up to see Carl Sagan walk into the room.

Perhaps the most famous living scientist of his day, Sagan had become known for his popular books and television series *Cosmos*, but he published in scientific journals as well. He proposed collaborating with the astonished young student on a research question: could Galileo, equipped with the usual instruments for studying planets, detect intelligent life on Earth? They poured over images with an eye for patterns, lines, and chemical signatures that mere geology could not explain. Nothing in Galileo's first encounter with Earth revealed that more than five billion people lived there, except for puzzling radio transmissions that were too scrambled to decode. Geissler at last found proof for intelligent life—"the hypothesis of last resort," as Sagan wrote—in a single image out of nearly six thousand snapped during the second flyby. It showed the straight-lined borders of national parks in Australia, the place Geissler still called home.

Galileo's journey had other surprises in store. In 1991, between the two Earth encounters, the spacecraft flew past Asteroid Gaspra, photographing the shadow-filled concavities of its pockmarked surface. Two years later, Galileo swung close to Ida, another main-belt asteroid drifting between the orbits of Mars and Jupiter. "This was a huge thing," Geissler said. "Nobody had ever seen one up close before. As far as anybody had ever seen, they were single points of light in a telescope. We had no idea what we were going to find."

Space in the returning data stream was painfully limited. After the first Earth encounter, Galileo's high-gain antenna had failed to open, forcing engineers to reroute the data through a smaller antenna. Images arrived in slices called "jail bars" that allowed the scientists to decide if it was worthwhile filling in the gaps. By putting these long, thin strips together, they could make out the shape of Asteroid Ida suspended gracefully against the empty backdrop of space. In one of the jail bars, a tiny white blip appeared above the asteroid.

"At the time, we weren't sure if we'd really discovered a moon of an asteroid," Geissler said. "We didn't even know if such a thing could happen." Data from the infrared spectrometer confirmed the finding. Little Dactyl hovered above the rotating asteroid, an unshakeable companion. Geissler printed color stereo images of the unusual pair, imaged in such spectacular detail that the rough, pitted surface of the asteroid stood out in sharp relief. That weekend, he pulled everyone who came near his house inside—from neighborhood kids to the mailman—to see the first three-dimensional view of an asteroid ever made.

HITCHING A RIDE

By this time, LPL resembled a kingdom of loosely connected fiefdoms made up of strong individuals leading separate research projects. "It's not the cohesive organization that it used to be," Floyd Herbert said with a hint of nostalgia, "but that's largely the result of success." As Levy had envisioned, spacecraft missions like Magellan, Galileo, and the Voyagers provided a profusion of new data that allowed researchers to chase their diverse interests. Robert Strom examined the tectonics of Ganymede, Carolyn Porco calculated the eccentric behavior of Uranus's and Neptune's rings, and Don Shemansky studied the roiling orange atmosphere of Titan. Armed with an intimate knowledge of ices, Uwe Fink pursued studies of comets.

Geology, astronomy, chemistry, physics, and atmospheric sciences blended in a mishmash of disciplines, bound together only by the shared space of the solar system. In the basement of the Kuiper Building, Bill Boynton examined pockets of calcium and aluminum within the famous Allende meteorite, thought to be the first solids condensed from the raw stuff of the solar system 4.5 billion years ago. David Wark, a postdoctoral associate in the meteorite lab, painstakingly scraped samples from the multilayered rims surrounding each inclusion.

Victor Baker took on geology students to study the morphology of Martian rivers, and John Lewis theorized about the kinetics that forge terrestrial planets. Many of the lab's original staff remained active in their early avocations: Elizabeth Roemer with comets, Tom Gehrels with Spacewatch, and Ewen Whitaker painstakingly correcting the error-strewn lunar nomenclature.

The members of Lyle's Garage, still housed in LPL West at the Gould-Simpson Building, continued their research by hitching a ride on the manned space program. Under NASA's Hitchhiker program, launched on the shuttle Endeavor in 1995, scientists could claim spare room in the shuttles for their instruments. The Arizona Airglow Experiment (GLO) flew multiple times with the Hitchhiker program, turning its ultraviolet gaze toward such picturesque targets as the Perseid meteor shower, Comet Hale-Bopp, and Earth's southern aurora.

Broadfoot's ultraviolet telescope, UVSTAR (Ultraviolet Spectrograph Telescope for Astronomical Research), also flew on seven shuttle missions to observe the Io plasma torus, the ring of charged particles emitting from Io's volcanoes that Broadfoot had discovered with his Voyager instrument. When the shuttle Discovery lifted off in 1998, astronaut John Glenn flew alongside both UVSTAR and GLO. That was UVSTAR's last flight. Galileo had arrived at Jupiter, providing much closer views than shuttles could offer.

Administrative staff made these scientific successes possible by handling the day-to-day requirements of the department. Lynn Lane, the senior business manager, started out as a secretary in 1978 at the age of twenty-three and built her career there, handling grant proposals and providing continuity to the ever-changing suite of projects passing through LPL. Lane estimated that LPL brings more revenue into Tucson than the University of Arizona's Wildcat basketball team. "You'll find a lot of us have been here for a very long time," she said. "I think that speaks for itself."

JUPITER'S FIREWORK SHOW

Only two years after the discovery of Chicxulub changed the way the world—from scientists to schoolchildren—thought about the demise of the dinosaurs, a discovery put Earth in the front row of an equally dramatic event. The night began inauspiciously enough. On March 23, 1993, Eugene and Carolyn Shoemaker drove to the Palomar Observatory in California with writer and astronomer David Levy to pursue their favorite pastime, hunting comets.

Legendary in the science community, the Shoemakers had formed the Astrogeology Research Program in Flagstaff at the dawn of the Space Age. A brilliant geologist, Gene Shoemaker had earned a master's degree from Caltech by the age of twenty and arrived in Arizona to hunt for uranium. Like

Kuiper, the Shoemakers recognized Arizona's potential as a focal point for planetary exploration. Now Gene and Carolyn went to Palomar Observatory seven nights a month to search for short-period comets. They had found eight so far with David Levy.

That night, with a dark weather system muting the lights of Los Angeles, the Shoemakers snapped off a few images through breaks in the clouds. They had decided not to waste good film on the grey skies and instead made use of a box of film that someone had accidently exposed to light. Jupiter, a black splotch in the negatives, marred the center of the field.

Two nights later, preparing for another evening of observing, Carolyn Shoemaker placed the developed Jupiter films beneath the stereomicroscope. A fuzzy streak appeared, trailing multiple tails. The oddly elongated object startled Carolyn. "This looks like a squashed comet!" she said. They needed confirmation. Levy knew that Jim Scotti, a member of LPL's Spacewatch program and longtime comet aficionado, was photographing the skies at Kitt Peak's 36-inch that night. A storm front had overwhelmed Palomar and headed toward Tucson, but Scotti agreed to look for the squashed comet before the desert skies clouded over.

Rain set in over Kitt Peak later that day, but not before Scotti confirmed the finding. The comet appeared as a sequence of nuclear fragments, each surrounded by a haze of evaporated particles. A few nights later, Wiesław Wisniewski joined Scotti on Kitt Peak to capture an image of the comet gracefully looping through the frame like a string of fiery pearls.

It soon became clear that Comet Shoemaker-Levy 9 was an unusual find. As the comet skimmed over Jupiter's cloud tops, the planet's gravity tore it to pieces. The best predictions revealed, to rising excitement, that the fragments would collide with Jupiter.

It was an unprecedented opportunity for scientists to view, from a safe distance, a type of event that could easily cause humankind's extinction. LPL hosted an informal conference in the broiling summer heat to coordinate observation plans. Nearly every major observatory on Earth, or orbiting above it, would be involved. Wisniewski was among those planning to participate in the global network of observers. He had turned his attention to near-Earth objects after his photometry work with Harold Johnson and had served in a similar global campaign to observe Gaspra and Ida during Galileo's close approaches.

It was not to be. The Polish astronomer died unexpectedly in February 1994 at the age of 62, leaving among his legacy the now-famous image of Shoemaker-Levy 9 en route to its final rendezvous. The first fragment slammed into Jupiter on July 16, 1994. Astronomers couldn't see the impact from Earth, but they noticed a fiery plume blossoming over the planet's curve just before Jupiter rotated the darkly scarred spots into view.

Sizzling fragments continued to bombard Jupiter's surface over the next six days. Steve Larson wasn't about to miss the solar system's greatest firework show. He joined the worldwide network of observers, sending custom-made spectrographs to observatories as far apart as Spain, Chile, Hawaii, and Australia. The observers, linked over the Internet, provided continuous coverage of the event as Jupiter set below the horizon in one area and rose in another.

The KAO provided even better views. The aircraft lifted off from Melbourne, Australia, making seven flights over the course of nine days. It flew at 41,000 feet, giving scientists a chance to make measurements without the interference of Earth's atmosphere. It was a last hurrah for the aging aircraft, which would retire the following year.

Two decades earlier, the KAO had discovered water vapor in Jupiter's atmosphere, hidden beneath the planet's ammonia-laden cirrus clouds. LPL researchers Don Hunten and Ann Sprague hoped that previously undetected chemical compounds would rise above Jupiter's sheltering clouds in the fireball caused by the comet's collision. Working with scientists from the NASA Ames Research Center, they installed an infrared spectrometer on the KAO telescope. The spectra revealed a spike in methane, ethane, and acetylene lit up by the heat of the collision, as well as sulfur dioxide, a compound not normally present in Jupiter's atmosphere.

Galileo had the best view of all. The spacecraft, still 140 million miles from Jupiter on its 2.8-billion-mile journey, had an unimpeded view of the luminous impacts as they occurred. The images, breathtaking in their drama, showed the bright fragments of Shoemaker-Levy 9 streaking toward the night side of Jupiter. Spectral emission lines tracked the fireball as it exploded above the cloud layers, providing the perfect complement to data obtained on the ground.

STANDING BACK TO SEE

Clearly, the solar system hadn't given up all of its secrets. When Galileo finally arrived at its destination in 1995, plenty had changed since the Voyagers passed through. Five months out from Jupiter, the spacecraft braved a barrage of dust streaming out of the Jovian system, possibly escaping from Io's volcanoes. The particle detector recorded dust levels up to sixty thousand times higher than normally encountered in the vacuum of space. On Arrival Day, December 7, Galileo looped Io and Europa and then settled into a close orbit around Jupiter, almost immediately detecting a previously unknown radiation belt.

LPL now had researchers equipped to work across the electromagnetic spectrum and with a wide range of instruments. Lyle Broadfoot and Bill Sandel joined Galileo's ultraviolet spectrometer team, eager for another look at the Io plasma torus. Don Hunten and Ann Sprague headed to Pasadena to contribute to spectroscopy studies. Martin Tomasko and his research associate Mark Lemmon worked on the Net Flux Radiometer experiment that rode Galileo's atmospheric probe into the maelstrom of Jupiter's clouds. The probe's data at once made obsolete the prevailing theories, finding more helium than expected and far less water.

For Rick Greenberg and his students, new images of Jupiter's moon Europa easily made up for the long wait. Greenberg had obtained a position on the Galileo Imaging Team by making the case that pictures could reveal details about the orbits and spins of satellites. Tidal forces distorted Europa's shape, stressing the icy surface into a crosshatched pattern of cracks. "We were able to explain some very distinctive crack patterns on Europa in terms of the tides," Greenberg said. "In fact, our explanation of a certain kind of crack pattern called cycloids was the first evidence that really said, 'There is liquid water there.'"

Studying the fault lines, Alyssa Sarid realized that the moon's frozen crust seemed to wander over what had to be a liquid interior heated from within. Sarid, Greg Hoppa, Randy Tufts, and Paul Geissler collaborated on a study with Greenberg that postulated an ever-changing surface where ice floes cycle beneath the surface, making room for new ones to form.

These features meant more than just a PhD dissertation for Tufts. His study of Europa's largest fault, the Astypalaea Linea, echoed his search for caves here on Earth. Its length rivals that of the San Andreas Fault in California. "It's simply a beautiful structure," Tufts told the *San Francisco Chronicle* that December. "I'd sure like to get to Europa to explore that fault."

As avidly as he campaigned for the preservation of Kartchner Caverns, Tufts now championed to protect Europa from contamination by spacecraft. Europa's liquid ocean intrigued scientists because it made the icy moon one of the most likely places beyond Earth to search for life. Tufts worried that Galileo, running low on propellant, might collide with Europa, breaking through the brittle crust and contaminating the moon with hitchhiking molecules from Earth.

In September 2003, NASA deliberately sacrificed Galileo to protect Europa, directing the spacecraft to plunge into Jupiter's crushing atmosphere. It was a fitting tribute to Tufts, who had passed away of a rare blood disease the year before. "It has beauty not because we are one with it, but because we are separated, standing back to see," Tufts once wrote in his journal. He was speaking of his beloved caves, but the words also ring true for Europa's startling seas.

SUNRISE AND SUNSET

After Viking, the world had to wait two decades to get another good look at the ochre surface of Earth's nearest neighbor. On July 4, 1997, Pathfinder bounced its way onto Mars, cushioned by airbags and cradling the tiny Sojourner rover like an egg. The Discovery-class mission was part of NASA's new "faster, better, cheaper" program, which centered on small, highly focused spacecraft. The joke circulating among engineers went "faster, better, cheaper—pick two," but Pathfinder nevertheless became a rousing success.

Peter Smith had won the NASA contract to build the imager for Mars Pathfinder in 1993. NASA's insistence on keeping the Pathfinder mission small and inexpensive led to a new opportunity for LPL because Smith's team would not only design the instrument, they would assemble and test it in Tucson, snapping practice photos of reddish volcanic rocks scattered haphazardly in the "Mars Garden" attached to the Kuiper Building.

"That changed my whole life," Smith said. "I had gotten into temporary work fifteen years earlier, and now I was a PI [principal investigator] of a $6 million project, and we had to design and build this camera, and I had a staff of thirty people. It was a very strange time for me. I had to grow into this new life very quickly."

On the Fourth of July holiday, LPL employees gathered at the Kuiper Building to celebrate. JPL reported the spacecraft's decreasing velocity during the four-and-a-half-minute descent. Smith waited. He was in the process of building an instrument for the next mission and writing a proposal for a third, trying to string together a career out of the risky business of space exploration. Wrapped in the petals of its descent vehicle, Pathfinder rolled to an undignified halt in Ares Vallis. Six hours later the sun rose on Mars, and Pathfinder opened its mechanical eye.

"That camera, I think, is what brought back the Mars program," Timothy Swindle said. The young professor had joined LPL in 1986, fresh out of

graduate school in St. Louis. He had settled on a specialty in Martian meteorites after wandering through bachelor's degrees in journalism, physics, and
math. Just before Pathfinder landed, he recalled, newspapers splashed headlines about the fiftieth anniversary of the Roswell incident. No one was paying
attention.

"Suddenly, the pictures came back from Pathfinder," Swindle said. "I remember sitting in the auditorium downstairs at the celebration for LPL employees,
and I had the family with me. All of a sudden here come these pictures from
the surface of Mars just flashing across the screen. No commentary or anything,
but real-time photos from the surface of Mars."

The world watched in delight as the twenty-five-pound Sojourner rover
navigated around jagged boulders. TV networks and newspapers carried new
images daily of the rock-strewn landscape, complete with dust devils, wispy
morning clouds, and rounded pebbles smoothed and tumbled by liquid water
billions of years ago. Over the course of its five-month mission, the Pathfinder
imager returned more than sixteen thousand images, including pale pink sun-
rises and distant, ghostly sunsets.

Pathfinder and Sojourner wouldn't be joined on the surface of Mars by
another working spacecraft until 2004, when Spirit and Opportunity bounced
down in their cumbersome airbags. The young flight director in charge of
bringing those rovers safely to the surface, Chris Lewicki, had been an aerospace engineering student at the University of Arizona and among the first
undergraduates involved in the Space Grant program. Susan Brew recalled
seeing him on the television driving the two rovers out of the wrappings of the
deflated airbags from Mission Control, "looking just like the student who had
walked through the door wanting to apply for an internship."

A string of failed and cancelled missions marred the intervening years, giving sleepless nights to scientists and engineers who had put their hopes on
Mars. But Pathfinder's final transmission in September 1997 did not leave Peter
Smith without a job. He had already begun planning the next big project—
capturing the unseen surface of Titan.

BENEATH THE CLOUDS OF TITAN

Titan's cloud-enshrouded surface had baffled scientists since Pioneer 11 first
attempted to peer through the thick haze. NASA planned to return to the

Saturn system with Cassini, an orbiter slated to arrive in 2004. It would drop an entry probe called Huygens into Titan's atmosphere, which would throw out a parachute and bump to a stop on the mysterious surface.

NASA officials had mulled over plans for Cassini-Huygens since the early eighties. The elaborate mission survived the subsequent budget cuts by virtue of sensitive political agreements with the European Space Agency. LPL played a prominent role in Cassini's planning and execution. Don Hunten helped formulate the concept of the mission. Carolyn Porco was the principal investigator of the Imaging Science Subsystem. Robert Brown led the Visible and Infrared Mapping Spectrometer Team. Newly arrived professor Roger Yelle got involved with the ion and neutral mass spectrometer experiment that would analyze Titan's atmosphere, and dozens of other LPL researchers found positions on the instrument teams.

Despite LPL's importance to Cassini, Martin Tomasko knew he was up against the odds. He wanted to propose an instrument called the Descent Imager/Spectral Radiometer (DISR) that would ride on the entry probe. A seven-part package of cameras and spectrometers, DISR would examine the opaque atmosphere and photograph the surface. If Huygens landed successfully, DISR's life span on the surface was expected to be all of twenty minutes.

Americans were designing most of Cassini's science experiments, but the European Space Agency had charge of the Huygens probe. Tomasko knew he needed to find some European partners if he wanted to make his proposal attractive.

"I remember tramping around France and Germany visiting various places and trying to draw out interest in being a collaborator on this experiment," Tomasko said. He was accompanied by Smith, who was helping Tomasko build the infrared spectrometers for DISR. Smith recalled his first glimpse of the castle-like dome of Paris Observatory—"totally different from American science; this was dripping with tradition and wonder."

Tomasko and Smith convinced a group of French scientists to help build DISR. They found another partner in Germany who agreed to provide the instrument's CCD detector. With little more than a handshake with the new collaborators, Tomasko wrote the proposal and won a $20 million contract in 1990. Collaborating with Martin Marietta, the aeronautics group in Denver that later evolved into Lockheed Martin, they had just a few short years to perfect the instrument before the looming launch deadline in 1997—all without crossing the Atlantic again.

Jonathan Lunine, a recent arrival on LPL's faculty, got a coveted slot on the three-member interdisciplinary science team for Huygens. A child of Apollo, Lunine remembered watching the first lunar landing from a hotel room at the Desert Inn Motel in Florida. "It seemed to me as a ten-year old it was the start of a new era," he said. He was right. Degrees in planetary science soon became available, and Lunine pursued his master's and PhD at Caltech.

As a graduate student, Lunine had made the bold suggestion that Titan might harbor a vast ocean of liquid hydrocarbons. After joining LPL in 1984, he became involved with the planning for Cassini, giving him a chance to prove his theory. Lunine won the bid to study Titan's surface as an interdisciplinary scientist. It was his first big proposal.

In 1997, the Cassini-Huygens spacecraft—an odd composite of computers, cameras, and communication dishes—launched from Cape Canaveral to begin its seven-year voyage. The carefully planned mission almost went awry when engineers discovered a glitch during the cruise. The Cassini orbiter, whistling overhead while Huygens dropped to the surface, was supposed to swivel around and relay transmissions from Huygens to Earth. But the orbiter's tremendous speed, compared to the parachuting probe, would create a significant Doppler shift. In other words, Cassini wouldn't hear the frequency on which Huygens transmitted.

Engineers had expected the Doppler shift, but somehow they had overlooked the necessary adjustments in the spacecraft's hardware. They rediscovered the problem when the European Space Agency ran a communications test a full three years after launch.

The solution, cobbled together by chastised American and European scientists, was elegant. The trajectory designers plotted a new tour that placed the orbiter at a much higher altitude above Titan, rearranging the geometry of Huygens's transmissions to Cassini to minimize the Doppler shift. To tweak the frequency further, they devoted one of Huygens's six batteries to baking the probe prior to its descent to Titan. The changes meant delaying delivery of the Huygens probe by seven weeks and tearing up Cassini's original schedule. It would take several months for the orbiter to resume its originally planned four-year tour of the Saturn system.

"It was only because the whole dang cruise was seven years long that they had enough time to do the test, discover this problem, and then come up with a solution," Tomasko said. "Otherwise, [Huygens] would've dropped into Titan's atmosphere and we wouldn't have heard a peep out of it, and nobody would have a clue why. It was a really close call . . . and a real story of some triumph that both the Europeans and the Americans can take credit for."

Cassini arrived at the Saturn system in July 2004. It settled into a looping, elliptical orbit around Saturn, occasionally passing by a moon for a closer look. In this complicated dance, Cassini skimmed close to Dione and Hyperion, plunged through Enceladus's icy plumes, and dipped low over Titan's orange haze, firing its thrusters to keep from tumbling into Saturn's pull—all while taking photographs from cameras directed, in part, by LPL personnel.

Huygens separated from Cassini on Christmas Day and hurtled into the opaque atmosphere of Titan three weeks later, on January 14, 2005. The investigators on the DISR team waited at mission control in Darmstadt, Germany. "Of course, everybody knew exactly what time the signals were supposed to arrive," Smith said. "You could tell that as the time came and went, the signals weren't arriving. Poor Marty was pacing."

The two antennas on the probe, A and B, could send redundant data for critical information in case one failed. That made sense for the complex spectral data collected by DISR, but Tomasko didn't want to waste precious space in the data stream by sending pictures twice. Each channel could transmit 350 images, and DISR would divvy up the images between the two for a grand total of 700. Cassini, orbiting above, was supposed to relay the data after the two-and-a-half-hour descent was complete. But the channel-A receiver didn't turn on.

As that realization swept through the room, everyone waited apprehensively to hear from the second channel. "I think six minutes went by, and then the signals started to come," Smith said. "In those six minutes, it was dying a thousand deaths."

The team in Germany waited in a tiny portable cabin as engineers extracted imaging data and sent it over to be deciphered. The young physicist at the computer screen, Bashar Rizk, had come to LPL's graduate program from Illinois. "I was consulting the catalog for [the University of] Arizona, and I turned one past the page for physics, and right after physics was planetary science," he said. "I stared at that page and thought, 'you've always loved astronomy and planets.'"

With an acceptance to the department in hand and an invitation to claim a room in Hawthorne House, Rizk threw everything he owned in his car and drove west of the Mississippi for the first time. After completing his graduate studies under Don Hunten, Rizk interviewed for a position on the DISR team. The tense moments ticking by in the portable cabin meant the culmination of years of research. Rizk called it "half a career for two hours of data."

As the imaging lead for DISR, Rizk had the privilege of the first look. "I said, 'Anybody want to see a picture of the surface of Titan?' and everybody leaped out of their chairs," Rizk recalled. "It was just a revelation seeing the surface beneath the clouds and the haze."

As with the first image of Mars three decades earlier, no one really knew what to expect. Converting the code on his computer, Rizk displayed the images as a set of thumbnails, one second per image. Lunine joined his colleagues in the cabin to see the pictures arrive.

> They were in random order, so you could see flashes of things on these thumbnails that looked like something you couldn't recognize, and then a river channel would pop up, and then something you couldn't recognize, and then a fracture would pop up. It was such a bizarre way to see them. Here were the first close-up images of Titan, and the last close-up images that we'll probably see for twenty years, all in that five-minute period. There was a lot of screaming in that room and I was one of the people screaming.

Every other image had been dropped out of the data, leaving the DISR team with the difficult task of making a mosaic out of a hodgepodge of pictures. To make matters worse, the team realized after their return to Tucson that the probe rotated in the wrong direction for nearly the entire two-and-a-half-hour plummet, bobbing and weaving unexpectedly in the atmosphere.

"Now you've got this set of 350 images you've got to put together," Tomasko said. "It's like a puzzle. You don't have a picture on the front of the box; you don't know what it's supposed to look like. The shape of the pieces change, half the pieces are missing, and the clock is going around backwards so you don't know where you are."

Beneath the obscuring clouds, DISR revealed a world of dark channels, dry lakes, and pebble-strewn gullies, with sand dunes and shorelines hauntingly similar to features on Earth. "There were pictures that looked like the coast of Italy," Smith said. "There were little rivers coming down off a hill into what looked like a lake. That's unbelievable. What a thrill. You don't know what to expect, but you don't expect rivers flowing down hills into lakes."

Its billowing parachute gathering scoopfuls of the nitrogen-rich smog, the Huygens probe plowed into sand and wobbled to a halt on the moon's surface. The frozen, organic-rich ground appeared dry, but liquid had clearly carved Titan's riverine features, perhaps in the form of methane raining down from the clouds or icy flows fissuring up from Titan's belly. Huygens continued transmitting data for an hour and a half, exceeding its predicted twenty-minute lifespan. Cassini's radar instrument later found the vast hydrocarbon lakes that Lunine had predicted, part of a global methane-ethane cycle that shapes the surface much as Earth's water cycle does, in average temperatures 180 degrees below freezing.

MAKING SPACE

The drought in space exploration had finally broken. Cassini-Huygens, a $3 billion Flagship-class mission boasting of an array of instruments and collaborators from around the world, achieved the Herculean task of bringing the space program back to the forefront. In the next decade, NASA's Discovery, Scout, and New Frontiers programs sent a host of new missions into space—"all short and sweet and different," as Tomasko described them.

These new opportunities allowed many LPL researchers and alumni to pursue cherished ambitions. Bill Boynton fulfilled a dream, and turned his luck around, when he joined the Near Earth Asteroid Rendezvous (NEAR) mission, later named NEAR Shoemaker, a Discovery-class probe that touched down on the precarious surface of Asteroid Eros in February 2001. Managed by the Applied Physics Laboratory at Johns Hopkins University, the spacecraft had orbited the asteroid for a full year before engineers decided to risk a landing. Boynton, serving on the science team for the X-ray/gamma ray spectrometer, suggested turning the instrument on after touchdown, and received seven days' worth of data about the asteroid's composition that far exceeded the quality of the data obtained from orbit.

Alan Binder still cherished the hope of walking on the Moon. Faced with the sting of America's disinterest in human space exploration, he did the next best thing. The 1998 Lunar Prospector mission, Binder said, was "the highlight of my career: I developed a spacecraft, engineered it, and flew it to the Moon." Lonnie Hood, who had once sifted through lunar data in Charles Sonett's lab, joined him to study the first new spacecraft data since the Apollo Age.

Begun as a private effort before the Discovery program made NASA funds available, Lunar Prospector set out to prove that good science could be done simply and cheaply. The desk-sized spacecraft cost "postage-stamp money," according to Binder, at a price tag of $63 million. It carried a suite of spectrometers, a magnetometer, and a gravity experiment. The spacecraft detected the small iron core that Hood had theorized, confirming its radius at no more than 450 kilometers, a perfect fit for the models that explained the Moon's origin by a giant impact.

Steve Larson, following Shoemaker-Levy 9's impact on Jupiter, turned his thoughts toward guarding the Earth from a similar catastrophe. He found avid supporters in graduate student Timothy Spahr and undergraduate Carl Hergenrother in the Department of Astronomy. Both offered to donate their time without pay to hunt for asteroids and comets.

"We were actually building our own computers from scratch because you couldn't buy a computer with the capacity," Larson said. "It was a very shoestring effort." After a slew of discoveries generated some attention from the media, however, they were able to purchase a prototype CCD camera to refurbish the Schmidt telescope on Mt. Bigelow. In 1998, they initiated the Catalina Sky Survey (CSS), a comprehensive search for near-Earth objects and potentially hazardous asteroids, a sister program to Spacewatch.

FIGURE 16. The Schmidt telescope on Mt. Bigelow, used for the Catalina Sky Survey, with the 61-inch Kuiper telescope behind

PHOTO BY AUTHOR

Over the next several years, the survey expanded to include two other telescopes, a 60-inch reflector on Mt. Lemmon and a 0.5-meter telescope in Siding Spring, Australia. Observers take multiple photographs of the same swath of sky, searching for objects that move rapidly against the predictable backdrop of stars. They post discoveries on the Internet, allowing hobbyist astronomers around the world to take up the task of tracking the object's orbit—a leap forward in technology that, in some ways, brought the Lunar Lab back to its origins as a place where amateurs could make important contributions to the field.

From its shoestring beginning, the survey expanded to employ eight full-time observers. One of those, Ed Beshore, had borrowed his dad's car in high school to make the two-hour drive to Lincoln, Nebraska, from his hometown, eager to see Bart Bok speak at the 1974 American Astronomical Society meeting. He remembered the famous astronomer as effusive and jovial. "I approached him to ask where I should go to become an astronomer," Beshore said. "He said, 'There's only one place you should go, and that's the University of Arizona.'"

Beshore took the advice. He graduated from the university's Department of Astronomy in 1977 and found a position on Tom Gehrels's Pioneer 11 team, writing software for the Saturn and Jupiter encounters. Beshore spent the next two decades as a professional engineer. When Larson began looking for computer-savvy staff to join CSS, Beshore jumped at the chance to return to Tucson and resume his first avocation. He joined CSS in 2002 and became its principal investigator in 2009.

CSS is now recognized as one of the most efficient surveys of its kind. Comet McNaught—"the most spectacular comet in our lifetime," Larson said—was discovered at Siding Spring in 2006. CSS scientists became the first to track an object on a collision course with Earth in 2008, discovering a car-sized meteorite just nineteen hours before it hit the ground, which enabled the recovery of fresh fragments from the Nubian Desert in Sudan. In fact, Larson estimated that two out of every three objects detected orbiting near Earth appeared first in their data.

Other ground-based projects found a home at LPL, complementing spacecraft missions. Asteroids, comets, and distant satellites were clearly rocky objects, and the skills required to understand them lay squarely in the realm of planetary science. Yet telescopes remained the only method of studying most distant objects, and rare, costly spacecraft missions required support from the ground.

"The ability to make reliable predictions of position is essential to plan spacecraft intercepts," Pat Roemer said. Her recovery of Comet Tempel 1 in

1967 and confirmation of its orbit in 1972 made it possible for the Discovery-class Deep Impact mission to intercept the comet in 2005, an achievement long in the making. Telescopes also enabled researchers to monitor dynamic processes such as Jupiter's Great Red Spot, impossible for short-lived space-craft missions. Astronomers and instrument-builders thus grew to rely on one another's data.

By 2004, the Lunar Lab employed three hundred people. Mike Drake fully expected to hire half that number again in the next several years as the lab pre-pared to support upcoming space missions. LPL took over the newly vacated Student Health Center across the street, christening it the "Charles P. Sonett Space Sciences Building" in honor of LPL's second director.

Lyle Broadfoot's research group moved to the Sonett Building from the ninth floor of Gould-Simpson. "The myth has always been, ever since we were down at Lyle's Garage on Ajo Way, that they're going to expand more space in the Lunar Lab main building, and then we'll be over there," Floyd Herbert laughed. "It's never happened and I'm sure it never will."

The scientists pitched in to renovate the old health center, and Broadfoot turned handyman to remove sinks from former examination rooms. Their research continued in the new surroundings. Broadfoot and Bill Sandel ana-lyzed data from a satellite called IMAGE. ("It's an acronym, and I'm really embarrassed by what it means, so I'm not going to tell you," Sandel said. The labyrinthine name turned out to be Imager for Magnetopause-to-Aurora Global Exploration.) Launched in 2000, IMAGE gathered data during its five-year mission about the magnetosphere, the region surrounding Earth con-trolled by its magnetic field.

Blasted by solar wind, the Earth's magnetosphere acts like a protective shield with a comet-like tail streaming out opposite from the Sun. The innermost part of the magnetosphere, called the plasmasphere, is a sloshing ring of tenu-ous plasma created by ionized particles in Earth's upper atmosphere. IMAGE crafted beautiful pictures of this dynamic region of near-space, invisible to the human eye except as the shifting light show of the aurora.

"With IMAGE, just coming to work every day has been a real gift because the plasmasphere changes all the time," Sandel said. "It's continually shrink-ing and expanding in response to forces from the outside, so you never know what it's going to look like the next time you look. . . . Our understanding has increased in big steps with each new mission. When you look with new eyes, you see new things. You can hardly avoid it."

FIGURE 17. The old Student Health Center was transformed into
the Charles P. Sonett Space Sciences Building in April 2004.

PHOTO BY AUTHOR

Broadfoot's group shared the Sonett Building with the Planetary Image
Research Laboratory (PIRL), which processed thousands of images from
spacecraft like Galileo and Cassini-Huygens. Robert Singer joined the lab in
1987 as PIRL's first director, working with a handful of graduate students to
painstakingly slice up and paste photographic hardcopies.

PIRL occupied a unique niche in the laboratory. Unfettered from any one
research project, it provided technological resources to scientists around the

country. Rather than reinventing the necessary computing resources for the many unique projects that passed through the Lunar Lab, PIRL offered a central place for top-of-the-line technology. Administrative and technical staff like Linda Hickcox and Bradford Castalia quietly carried the organization forward, providing continuity and support as spacecraft missions came and went.

In 1996, Alfred McEwen became PIRL's director. Under his leadership, PIRL processed thousands of images from the Cassini-Huygens mission. Around the same time, the opportunity arose to lead the High Resolution Imaging Science Experiment (HiRISE) on board the Mars Reconnaissance Orbiter (MRO). McEwen knew he had to take it. A geologist, he had come to Arizona with an eye for the bare rocks and water-formed features so easy to see in the grassless landscape. Mars, with its bleak horizons, was irresistible.

McEwen teamed up with Alan Delamere of Ball Aerospace to build the bulky, 145-pound instrument. It launched aboard MRO in August 2005. "To me, the most exciting moment was getting our very first images in-flight, pictures of the Moon and some stars," McEwen said. "They're not exciting images—we were a long ways away from the Moon—but I knew what it meant. I knew that it meant our camera was working."

The following year, HiRISE pointed its three-meter, high-gain antenna toward Earth to beam back unprecedented views of the Martian landscape. HiRISE photographed a dynamic planet, tracking the motion of seasonal frost expanding and contracting, dust showering down the slip face of a dune, and graceful arcs of sand migrating slowly across desolate plains. The camera showed Mars with an artistry that rivaled the fantastic dreamscapes of the most imaginative paintings. Later it would capture family portraits of Spirit, Opportunity, Phoenix, and Curiosity as those spacecraft explored the surface. "MRO has already returned more Mars data than all the previous Mars missions combined," McEwen said. "It's a huge amount of data, and it's very high-quality data for science. This is really the scientist's mission."

AT THE EDGE

The careworn Voyager spacecraft still illuminate the unknown edges of the solar system. Two decades after ending their planetary tours, they began their plunge out of the heliosphere, a bubble surrounding the solar system carved out in interstellar space by energetic particles streaming from the Sun. At this distant frontier, solar wind piles up against a flimsy dam of interstellar gas, the thin material that fills the space between stars.

Voyager 1 first touched the edge of the heliosphere in 2004 when it crossed the "termination shock," the place where solar wind particles, pressing into the interstellar medium, slow down to the speed of sound. Three years later, Voyager 2 made multiple crossings into the termination shock. Everyone awaited the moment when Voyager 1, leading the way, would leave the heliosphere and cross into interstellar space—but it proved surprisingly difficult to pin down a date.

NASA eventually concluded that the crossing had occurred in August 2012. They were helped by unexpected outbursts from the sun, which sent particles hurtling outward on a four-hundred-day journey to the edge of the heliosphere. When these outbursts plowed into interstellar space, Voyager 1 felt the tremors: imagine a wave crashing into the shore, rocking a child's sailboat bobbing on the tideline. The vibrations allowed scientists to measure the density of the surrounding plasma. It was forty times higher than previous measurements, a sign that Voyager 1 had reached interstellar space.

It's up to solar physicists such as Randy Jokipii and Joe Giacalone to interpret the often confusing data still echoing back from the Voyagers. "There's nothing like observing new things," Jokipii said. "I think most scientists feel that way, no matter what their field is. We're right at the farthest we've ever been, and we've found new things out there."

Voyager 1 isn't completely out of the heliosphere. The Sun's influence still marks the surrounding plasma, much as a tide tosses occasional debris up onto the shore. In roughly a decade, the last of the spacecraft's instruments will shut down and Voyager 1 will go silent. In three hundred years, it might reach the Oort Cloud, often considered to be the true edge of the solar system. In forty thousand years, it will have reached the midway point between the Sun and the next nearest star.

For now, the ultraviolet spectrometers built by Broadfoot's research group still function, staring in one direction as the Voyagers drift over the edge. At the extreme end of the ultraviolet spectrum, the instruments catch glimpses of interstellar gas as it absorbs the last glimmers of sunlight. Lyle's Garage is now just a memory. Its remaining members moved into other research, allowing a new generation of faculty and students to inherit the euphoria of spacecraft missions. "The excitement of the old days has passed on to the younger generation," Herbert said with a hint of envy. "The rest of us are all basking in our old glory days."

THE OTHER SIDE OF MERCURY

At first glance, Mercury had appeared as barren as Earth's moon, a charred, dense ball of rock with an iron core, scarred by impact craters and void of an atmosphere, its sky stretching pure black from horizon to horizon. In full sunlight, Mercury's surface skyrockets to eight times the temperature of Tucson's worst summer day. It seemed an unlikely place to find water ice and organic compounds, the combination that scientists sought so diligently on Mars.

Mariner 10, the first spacecraft to visit Mercury, lifted off from Cape Canaveral a month before Kuiper's death. It imaged almost half the planet's surface before its maneuvering propellant was depleted. Sailboat-like, the spacecraft made one final loop around the Sun with its panels spread to the solar wind before running out of gas.

"It wasn't supposed to be a definitive mission," Robert Strom said. "Reconnaissance was all it was." Strom served on a planning committee that intended to follow up with an orbiter, thrilled by the planet's geology and eager to see the other side. But the fervor of Apollo had passed, and interest waned. Strom, now an emeritus professor and one of the last remaining early members of the Lunar Lab, wouldn't see Mercury again for more than thirty years.

In 2004 NASA launched MESSENGER, a Discovery-class mission equipped with instruments to study Mercury's geology, chemistry, and magnetosphere. (The mission's full name is "MErcury Surface, Space ENvironment, GEochemistry and Ranging," possibly the most torturous example of NASA's love of acronyms.) Four years later the spacecraft completed its first Mercury flyby. Until MESSENGER, Mercury remained the least-known planet in the solar system, forgotten in the years since the Space Age. "When I saw those first images coming back, it was extremely emotional," said Strom, who served on the science team as a geologist. "I had tears in my eyes. I'd waited so long."

Two more flybys successively slowed the spacecraft in preparation for entering orbit in 2011. By the second flyby, the combined images—new and old—covered 90 percent of the planet's surface. At long last, Strom saw Mercury's far side. The planet has almost no tilt, leaving the flat floors of craters on the poles in permanent shadows. Without an atmosphere to hold in heat, Mercury's surface drops to -170°C in the darkness.

Astronomers had earlier noticed bright patches at Mercury's poles consistent with water ice. Many of these patches correlated with large impact craters mapped by Mariner 10. With less than half the planet mapped, however, the water ice hypothesis remained uncertain—that is, until MESSENGER's arrival. Its neutron spectrometer detected nearly pure water ice rimming shadowed craters. More ice was buried beneath an insulating layer of mysterious dark material in slightly warmer regions. Scientists speculated that the dark sheen might indicate organic compounds, brought to Mercury by comets and asteroids that pummeled the planet.

Mercury earned a new impact crater in April 2015 when MESSENGER ran out of propellant and crashed. Strom's passion for overlooked Mercury proved worthwhile. The hot little planet had shown that organic molecules, the precursors of life, might be found on the most unlikely of worlds. In Mercury and other terrestrial planets Strom found examples of the myriad ways a planet can form and, over the course of time, completely remake its climate and terrain.

In the early nineties, Strom taught a course at the University of Arizona on planetary catastrophes, covering everything from volcanism on Io to flooding on Mars to the cataclysmic resurfacing of Venus. Almost as an afterthought, he added climate change to the syllabus. Scientists had only just begun to voice concerns about Earth's rising global temperatures. Strom began researching the subject full-time after his retirement.

Five years and three hundred pages later, he had written a book explaining the irreversible experiment that humans have undertaken with our planet. An early chapter in *Hot House* describes the lessons learned from Earth's nearest neighbors, Venus and Mars, which have both experienced dramatic changes in their history. At an extreme scenario, Strom said, climate change "could easily lead to the end of civilization as we know it."

Despite that sobering knowledge, Strom remained an optimistic and enthusiastic teacher. Almost two decades after first exploring the subject, he taught another freshman seminar, this one focused solely on climate change. "Some of the students that I deal with, they're great, wonderful people and extremely smart," he said. "To lose that would just be awful, just a tragedy that I don't even want to think about. So that's why I wrote the book. That's my sayonara."

SAVE THE METEORITES

Half a century after the Lunar Lab's founding, the widely ranging interests of LPL's faculty might have surprised Kuiper's early staff. No longer crowded into Quonset huts, LPL outgrew its boundaries both physically and intellectually. Mike Drake, who had now held the director's chair longer than anyone else in the lab's history, wasn't concerned particularly with the interests of the faculty he hired as much as their talents and imaginations. "One of the things I learned long ago is not to dictate how people do things," Drake said. "Create the opportunities for them to explore, and you'll be astonished at what they come up with."

The field of planetary science continued to fracture into a multitude of disciplines, ideas, and approaches. Theorists found a place at LPL alongside of engineers and experimenters. Victor Baker characterized the geomorphology of planets by studying the detailed, sweeping vistas returned by close-approaching spacecraft. Adam Showman traced the changing patterns of Jupiter's banded clouds with data from the Pioneer days. Other researchers looked beyond the visible wavelengths: Joe Giacalone studied the dynamic cycle of spots that dance over the surface of the Sun, and Jozsef Kota researched the cosmic rays that fill the emptiness between planets. Astronomers still found a place at LPL, such as Jay Holberg, who contemplated the death of stars with his research on faint white dwarfs that slowly cool in the sky.

Comets, asteroids, and meteorites remained objects of interest, hovering in the gray area between planetary science and astronomy. Tim Swindle continued to hunt for clues to the ages and turbulent histories of meteorites. He had set up his first lab on the fifth floor of the Kuiper Building in the late eighties, swinging the expensive equipment through the window with a crane. Renu Malhotra, a theorist, studied objects in the distant Kuiper Belt, modeling how they migrate from the far edge of the solar system to bombard the surfaces of planets.

Drake himself, though busy with the difficult task of coordinating communication between the far-flung corners of the department, continued his research on the origin and evolution of the Moon, asteroids, and planets. Born in England, Drake relocated to the west coast of America at the age of twenty-one, searching out a sunny place to study geology with the idea of becoming an oceanographer. Apollo astronauts landed on the Moon just two years later, and Drake published his first paper on Moon rocks in *Science* a year after that.

At LPL, Drake's research involved a microscopic scrutiny of metal-seeking elements such as nickel, cobalt, phosphorus, and gold, but he always kept in sight the big questions—the origin of the solar system and the conditions that spark life. In the basement of the Kuiper Building, Drake set up an electron microprobe lab equipped to examine tiny slices of meteorites, overseen first by Tom

Teska and later by Ken Domanik. He ran complex experiments to simulate the metal core formation of the Moon and asteroid 4 Vesta, and then turned his thoughts to Earth, hypothesizing an ocean of magma in the planet's early history.

Drake had attended the 1984 Kona conference on the origin of the Moon but had started out as a skeptic of the giant impact theory, in part because the Earth, unlike the Moon, showed no evidence of a molten beginning. In the early nineties, however, improved laboratory techniques made it possible for Drake to match the concentrations of siderophile elements in the Earth's mantle with predictions for a deep magma ocean. That led him to a new theory for how primordial water arrived on Earth, as molecules so tightly attached to cosmic dust grains that they survived the metal-melting temperatures of the planet's violent birth.

Meanwhile, Drake had a laboratory and academic department to run, no small task for a place that had grown so large and diverse. "I chose the guiding principal that the best research department will be the best teaching department and the best outreach department," Drake said. "Every decision I've made has been examined in that mirror." He upheld the high standards for education that Eugene Levy had set, right down to his own research projects. Students remembered how cheerfully he made time to listen to their ideas and offer guidance.

The investment paid off in a new generation of planetary scientists. "One of the best things you can do is provide opportunities for undergraduates to get involved in spacecraft missions," Dante Lauretta said. "My story shows that." Lauretta entered his fifth year at the University of Arizona without knowing what he wanted to do next. He had spent the summer working as a cook and bartender in California and living in a 1972 Volkswagen Bus, and would soon graduate with three bachelor's degrees in mathematics, physics, and Japanese. All he knew for sure was that he didn't want to work the early-morning breakfast shift on weekends anymore.

Then he saw an advertisement in the *Arizona Daily Wildcat* for the Space Grant program. He hadn't even known that LPL existed, but the idea caught his imagination. He applied, and received the fantastical assignment of designing a mathematics-based language to communicate with extraterrestrial life. Less than a decade later he joined the faculty at LPL to study the formation of planets and the origin of life.

Lauretta sold his VW Bus and moved to St. Louis for graduate school, intending to get involved with Mars Observer. Its failure sent him into more dependable laboratory research, vowing to never get involved with spacecraft again. At LPL he applied for NASA funding to study phosphorus, a chemical element critical to DNA, RNA, and the basic metabolism of cells. He theorized that meteorites seeded the early Earth with the building blocks for life, and he enlisted the help of graduate student Matthew Pasek to show that iron meteorites, sizzling into freshwater lakes, could spark the synthesis of critical phosphorus molecules.

"It was one of those rare things in science where you hit the jackpot almost right from the start," Lauretta said. He began hunting for meteorite samples at Tucson's Gem and Mineral Show, an annual event. Among the closely spaced rows of specimens for sale—fossils in glass cases, minerals jumbled in bins—Lauretta met Marvin Killgore, a meteorite dealer with a passion for scientific inquiry. The former plumber, miner, and gold prospector had turned his hobby into a career, becoming a self-taught expert in geochemistry.

Lauretta and Killgore realized they faced a common problem. "The commercial meteorite world was in a crazy state," Lauretta said, "with meteorites being harvested at very rapid time-scales compared to the rate at which they fall on Earth. This was kind of a bonanza, similar to the gold rush period in the 1800s, and it was going to die out like the Gold Rush did. All the good meteorites were going to be recovered, and if we didn't do something to preserve them, there'd be nothing left for future generations to work on."

Killgore's private collection lay scattered in storage facilities and research institutions around the country, and LPL, once again, had run short of space. Peter Smith had won the contract for the Phoenix Mars Scout mission, and Drake needed a place to house the scientists and engineers that would soon flock to Tucson.

Drake and Smith convinced the university to purchase a building south of campus, and in May 2004 the Lunar Lab expanded into the Phoenix Science Operations Center. The place would soon predominate in Lauretta's career, but he didn't know that yet. More immediately, it meant a home for the Southwest Meteorite Center, the organization Lauretta and Killgore created to preserve meteorite specimens for future generations to study.

That year, with the launch of Phoenix still a gleam on the horizon, Drake approached Lauretta with a scheme for a new mission. He wanted to draft a proposal to NASA for a decades-long project to send a robotic visitor to an asteroid and bring a sample home. Up until now, Lauretta had kept to his renouncement of spacecraft missions. Yet his laboratory research, centering on fundamental questions about water, organic molecules, and life among the stars, had exactly prepared him for this chance. "I had it all right there at my fingertips," he said. They buckled down to the task of convincing NASA they could pull off the ambitious plan.

First, however, LPL had to prove that it could do more than just contribute instruments and ideas to NASA. The little laboratory that Kuiper had founded fifty years before now aspired to run a mission of its own.

FIRE AND ICE

Despite the crushing loss of the Mars Observer in 1993, Bill Boynton hadn't given up on sending a gamma ray spectrometer to Mars. Engineers revamped spare instruments from the failed mission for two more orbiters. Mars Global Surveyor arrived at its destination in 1997. Four years later, Mars Odyssey began circling above the surface of the Red Planet. It carried the fulfillment of a dream: Boynton had designed and built a gamma ray spectrometer for the mission, putting the pieces together in his lab at the Kuiper Building.

Peter Smith was also waiting for a second chance. In 1999 the Mars Polar Lander had fallen to an unknown fate over the southern hemisphere, taking with it Smith's imaging system and Boynton's Thermal and Evolved Gas Analyzer (TEGA), a set of tiny ovens that would have sniffed out the chemistry of Martian soil. The highly publicized failure brought the Mars program to an inglorious halt, and NASA cancelled the 2001 Mars Surveyor lander. Smith turned his attention to Beagle 2, building a microscope for the European lander with a small LPL team. It also failed. HiRISE later photographed the lander, intact on the surface but without its solar panels fully deployed, which apparently rendered it silent. To Smith, it was beginning to look like a career in planetary science wasn't such a good choice after all.

Boynton's gamma ray spectrometer aboard Mars Odyssey changed everything. "We discovered vast quantities of ice buried just beneath the surface that nobody knew was there. That really changed people's thinking about Mars," Boynton said. As radiation streams down onto Mars's unprotected surface, the regolith releases energetic particles that carry the fingerprint of the elements they struck. Boynton's instrument, designed for both gamma rays and neutrons, detected a strong signature of hydrogen hidden in the top three feet of the stony surface. It was the first clear proof that water existed on Mars in the present time.

With untarnished optimism, Smith grasped the chance. Hardware for the partially built Mars Surveyor waited in a clean room at Lockheed Martin, and several instrument-builders from Mars Polar Lander were ready to try again. Smith wrote a proposal for NASA's first Scout-class mission, keeping it within the tight constraints of the budget by reusing and improving old parts. Engineering a new mission out of the ashes, Smith named the spacecraft Phoenix.

"To me, that was it," Smith said. "NASA's theme is 'follow the water,' and no spacecraft had ever gotten anywhere close to water. Where they had landed, there hadn't been water for three billion years. I thought: here's a chance. You could just land anywhere in the polar region, and ice is under you. You don't need wheels. Our mission is vertical."

The Mars Reconnaissance Orbiter made the daunting task of choosing a safe and scientifically interesting landing site much easier. Shortly after LPL scientists switched on HiRISE in the autumn of 2006, new photos revealed that car-sized boulders lay strewn across the primary choice—their Halloween gift to the Phoenix team, Alfred McEwen joked.

The team had to choose a new site, and Smith finally settled on Vastitas Borealis. The broad valley floor had fractured into polygonal shapes, much like the arctic tundra of Earth, and Smith took it for a good sign. In Antarctica, similarly patterned terrain signals the presence of ice beneath the soil, expanding and contracting during freeze-thaw cycles.

As work on Phoenix progressed, HiRISE photographed the site under changing seasons, watching frost sweep over the surface in the Martian autumn and fade away each spring. "Phoenix is looking at what's happening today," Smith said. "Ice is not ancient. We want to look at modern structures and modern processes. Is there any chance the ice did melt, and if so, was there biology? All of a sudden, it clicked. I've been awfully lucky. Things have clicked several times for me. You're lucky enough to have that happen once or twice in your life."

Scientists and engineers from all over the world came to Tucson to work on the mission in the blistering summer. "When we wrote the proposal, one thing I insisted on against tremendous resistance was doing the operations here," Smith said. "I convinced the university to agree to provide us space. My feeling was the science expertise is at the universities. The students are here. We've got access to all kinds of resources that scientists need."

The Phoenix lander, supported by a sturdy tripod and powered by two circular solar arrays, had an eight-foot robotic arm with a tiny rasp to scrape up samples of the icy regolith. Boynton's improved version of the TEGA instrument would heat the samples, marking the temperature of phase transitions as various materials vaporized and sending the gases to a mass spectrometer for analysis. Phoenix also included a tiny chemistry laboratory, a microscope, a weather station, and two cameras, one attached to the robotic arm and another to serve as the mission's "eyes" and return sweeping panoramic views of the arctic plains.

On May 25, 2008, HiRISE turned its eye toward Heimdall Crater, snapping off a shot. Phoenix plummeted toward the surface, its parachute streaming out behind. Never before could one spacecraft photograph another on its descent to a planet's surface. At JPL, Richard Kornfeld reported on the entry, descent, and landing sequence in his calming voice as the telemetry came in. Parachute deployed. Heat shield triggered. Landing legs deployed. Then came the countdown of altitude as the spacecraft dropped from its backshell at 1,100 meters in freefall. Moments later, Phoenix fired its reverse thrusters and jolted to a halt on the surface of Mars.

With Phoenix safely on the ground, JPL handed the reins to Tucson. Mike Drake, watching the big screens in the Science Operations Center surrounded by colleagues, staff, and students, called it a seminal moment. "It's the first time a major space mission has been controlled by a university once it's on the surface of the planet," he said. Within hours, Phoenix returned a triumphant first photo of its footpad planted firmly on the surface of Mars.

Over the next 151 Martian sols—a little more than five months on Earth—Phoenix confirmed the presence of water ice beneath the soil of the artic plains and sampled the alkaline soil with its cocktail of calcium carbonate, perchlorate, and salt. Its instruments detected snowfall in the pink upper atmosphere, leaving wispy contrails that bore a striking resemblance to the vaporizing rain that streaks Tucson's sky during monsoon summer. The findings suggested a Mars with ongoing climate cycles—a Mars that might tilt into favorable conditions for liquid films of water clinging to the soil grains, with the help of perchlorate to lower the freezing point.

That was all still ahead. On Landing Day, crowds numbering in the thousands—students and scientists standing side by side—waited at the Kuiper, Sonett, and Drake Buildings (as the Science Operations Center would soon be called) for the first images to return. Outside, evening fell. Mars gleamed over the western horizon, a bright speck in the cloudless sky.

FIGURE 18. The Phoenix Mars Scout Mission team assembles outside the Science Operations Center

COURTESY OF LPL SPACE IMAGERY CENTER

A PIECE OF A STAR

The Science Operations Center brought the number of premises LPL had occupied over the last fifty years up to ten, a far cry from the Quonset huts where everything began. It became a favorite landmark in the Tucson community as well as a center for research, housing a double of the Phoenix lander and bearing a fiery mural on its south wall depicting the journey from Earth to Mars. Secretly, many LPL scientists hoped for a seamless transition from one mission to the next within the building's walls. Phoenix couldn't last forever, limited by its solar panels and destined to disappear beneath the crushing weight of carbon dioxide ice in the Martian winter.

But by the time of Phoenix's final transmission on November 2, 2008, Drake and Lauretta had failed twice to win a NASA contract for the asteroid sample return mission. "Mike and I were both really green," Lauretta recalled. "We didn't know what we were getting into. We didn't know what it took to lead a mission." LPL had a longstanding reputation for good science and a strong aerospace partner in Lockheed Martin, but it lacked skill in mission management, which required expertise in finance, business, and law as well as science and engineering.

Drake and Lauretta submitted the proposal for the second time with an additional partner, NASA's Goddard Space Flight Center, to handle management. They spent a heartbreaking year in the final round of the competition developing the details of the complex mission, only to fail again. They simply could not afford an asteroid sample return mission on a Discovery-class budget.

Then Phoenix landed on Mars, and Tucson blossomed once more into national prominence as a leader in space exploration. The same year, the National Research Council released its recommendations for NASA's New Frontiers Program, which offered twice the budget of a Discovery-class mission. A sample return from an organic-rich asteroid made the list. "That was not an accident," Lauretta said. "We had done a lot of work communicating to our peers in the science community the value of this project. We had made an impression."

OSIRIS-REx, as the mission is now called, would seek answers to the big questions that had driven the careers of many planetary scientists. As Drake told reporters, the mission centered on humankind's origin and destiny. Precise measurements of an asteroid's orbit would help improve ground-based tracking of near-Earth objects, a task that had taken on some sense of urgency since the dramatic example of Shoemaker-Levy 9. Moreover, the spacecraft would return a sample from a pristine remnant of the solar system's chaotic birth. Scientists hoped to find its surface chock-full of carbon and volatile organic compounds.

"When you look at carbon, it's wrapped up in intriguing organic molecules like amino acids, nucleic acids, and sugars—all the ingredients for life," Lauretta said. "It's an incredible hypothesis that the Earth is habitable because the ingredients were delivered by asteroids." A sample from the asteroid's gravelly surface would offer a glimpse back in time to the start of it all—a glimpse relatively unfettered from the uncertainties that plague meteorites, which Drake called "free space missions" that arrive on Earth much altered by their journey.

On May 25, 2011, exactly three years after the Phoenix landing propelled LPL into its untested role as the leader of a spacecraft mission, Drake received word that NASA had selected OSIRIS-REx. "That will be my first personal experience running a space mission, and it's going to be exciting," said Drake, who hadn't worked with a spacecraft sample since the Moon rocks in his graduate school days. "Returning samples is the Holy Grail for almost any scientist. Even an astronomer would sooner have a piece of the star than look at the light emitted from it."

But Drake did not live to see the mission launch. He died only four months later, and Lauretta, the young scientist he had handpicked as a partner, stepped into the role of principal investigator of the OSIRIS-REx mission. The half-emptied halls of the Science Operations Center sprang to life, its offices filling once more with scientists at all stages in their careers. "Mike was very clear he had every confidence in me," Lauretta said. "We had a lot of time to talk about it. In many ways I'm doing this to honor him and keep his legacy alive." In November 2011, the Science Operations Center received its final name, the Michael J. Drake Building.

THE SEEDS OF LIFE

Two events occurred during the selection of OSIRIS-REx that had bearing on the new mission. The first took place on the night of October 6, 2008, when a CSS scientist named Richard Kowalski tracked a near-Earth object—a *very* near-Earth object—from the observatory on Mt. Lemmon. More than two dozen observers took up the watch as 2008 TC_3 plunged into the Earth's shadow and vanished. Nineteen hours after discovery, it exploded over the bleak horizons of the Nubian Desert in a brilliant fireball.

The fragments, recovered two months later, revealed something surprising: more than 99 percent of the original object had vanished in the descent. The pieces that remained harbored the cooked remains of amino acids, the code that maps out proteins in all known living things.

Scientists had long conjectured that asteroids might have rained down water and organic molecules onto early Earth, sparking its journey toward a habitable planet after its fiery inception. However, as 2008 TC$_3$ proved, only a sliver of fragile meteorites survived their roller-coaster rides through the atmosphere. They made poor representatives of untouched asteroids slinging through space in their orbits. OSIRIS-REx therefore had hinged part of its mission objectives on a question: did asteroids actually harbor the seeds of life?

In February 2010, the hopeful members of the OSIRIS-REx team held a meeting in Houston. Among the guests were two LPL alumni, Humberto Campins and Josh Emery. The conversation turned to the perplexing issue of selling a sample return mission to NASA when water ice and organic molecules had not yet been detected on an asteroid.

Campins, unable to keep silent, raised his hand and said, "Yes, they have!"

The news made a fireball of its own in the scientific world. With colleagues at the University of Florida, Campins was studying an asteroid called 24 Themis. He wasn't looking for water ice. Water slowly sublimates even in the coldness of space, and a main-belt asteroid with several billion years' worth of birthdays could hardly be expected to have any.

Yet the infrared spectra of the asteroid's surface showed intriguing absorption features. Campins called Emery and Andrew Rivkin, another LPL graduate and former resident of Hawthorne House, who were studying 24 Themis from the same observatory on Mauna Kea.

"I contacted them and said, 'We have this beautiful spectrum that we can only fit with water ice,' and they said, 'Yes, we agree,'" Campins recalled. "Rather than trying to scoop each other, I proposed that we write and submit the papers at the same time."

Later, Campins sent his new partners an email noting that he saw something in the spectrum that looked like carbon-hydrogen bonds. They never responded, and Campins forgot about the incident as he focused on analyzing the water ice. Just before the two teams planned to submit their papers to *Nature*, Emery and Rivkin rediscovered the signature in their own data.

"We had a week to change the paper," Campins said. "This combination of water ice and organic molecules, the two most important ingredients for our origin of life, had such an impact."

They couldn't give any details at the Houston meeting because the information was embargoed until its official publication that April. Together, the two research teams had analyzed the entire surface of the asteroid, proving their discovery wasn't a chance accident. The prevalence of water ice remained a puzzle: the scientists guessed that some subsurface reservoir continually pushed new material to the surface. The seeds of life had been found.

A THOUSAND DAYS OF DISCOVERY

Everything had fallen into place for Tucson once again to launch into a leadership role in space exploration. A vital task remained: to choose the asteroid that would dominate the careers of several hundred scientists and engineers for the next fourteen years.

During the development of the OSIRIS-REx proposal, Lauretta approached Carl Hergenrother, a former Space Grant student who had helped found CSS and now had his bachelor's degree in astronomy, and offered him a position on the team. Hergenrother dropped his vague plans for graduate school and accepted at once.

Drake and Lauretta had settled on a prime target for the mission, but closer observations revealed that the asteroid wasn't carbonaceous. They wanted a B-type asteroid, a rare subgroup of C-type (carbon-rich) asteroids thought to be relatively unchanged since their formation. No carbonaceous asteroid had ever been directly sampled; thus far, scientists had to rely on meteorites and spectral characteristics to learn about their origin and composition.

Hergenrother started looking for a new target with Kuiper's venerable 61-inch telescope in the Catalinas. He quickly realized that 1999 RQ36 (later named Bennu) was a perfect fit: a dark B-type object, about 575 meters wide, that veered close to Earth every six years. "I emailed Mike and Dante pretty much from the mountain," he said, "and within two or three days it had become our target. We actually had an image of it, which you don't have for most asteroids."

By this time, generations of LPL graduates had scattered across the world, many of them becoming eminent scientists in competing institutions or even founding new ones. OSIRIS-REx meant a reunion for alumni who earned a place on the mission's roster. Campins and Emery both joined the team to lend their expertise in spectroscopy. Alan Hildebrand led the Canadian team that built the laser instrument that would create a global topographical map of the asteroid. Bill Bottke headed up the Dynamical Evolution Working Group, tasked with understanding Bennu's orbital history. Bashar Rizk found a place on the camera team, and Ed Beshore relinquished the directorship of CSS to become the deputy principal investigator. Team members ranged from experienced spacecraft scientists like Bill Boynton to brand-new Space Grant interns.

As head of the Asteroid Astronomy Working Group, Hergenrother's job is to learn as much as possible about Bennu and other NEOs from ground-based telescopes. He enlisted the help of an oft-overlooked resource: the public. Amateur astronomers, unburdened with the time and funding constraints that often plague professionals, can make vital contributions to the tedious work of finding and tracking solar system objects.

The Target Asteroids! program asks amateurs to track NEOs with their own telescopes or borrowed equipment, measuring their position, brightness, and spectra. Hergenrother compiled a list of objects more than two hundred meters in size—large enough to flatten a city—for amateurs to pursue. He leads the program with geologist Anna Spitz, the education and public outreach lead, and meteorite expert Dolores Hill, who came to LPL to work in Bill Boynton's lab in the early eighties.

Bennu, of course, takes the spotlight in this widespread effort to characterize NEOs. OSIRIS-REx will return the favor by "ground-truthing" the data collected from telescopes or inferred from meteorites. Not only will the mission shed light on the intriguing question of life's origin and distribution in the universe, but it will help safeguard Earth against the catastrophes that scientists now know formed the Moon and extinguished the dinosaurs.

Bennu has a one-in-eighteen-hundred chance of striking Earth in the twenty-second century, high by the vast standards of space. Sunlight absorbed by the dark surface and radiated out on the hot afternoon side acts like a tiny reverse thruster, nudging the asteroid in unpredictable ways. By measuring this phenomenon, called a "Yarkovsky effect," OSIRIS-REx could improve the tracking of thousands of NEOs.

"Eventually, astronomy is physics, but planetary science is exploration," Spitz explained. "Are the assumptions we make from the ground correct? You don't know unless you go there."

Perhaps the most remarkable feature of the OSIRIS-REx mission is its length, spanning fourteen years from the proposal's acceptance to the sample return, followed by years of analysis. The spacecraft is scheduled to launch in 2016 and approach Bennu in the fall of 2018. Mission engineers will command the spacecraft to photograph and map the asteroid (in visible, infrared, and X-ray wavelengths) from a distance of 400 million miles.

OSIRIS-REx will carry spectrometers, sampling equipment, and three high-resolution cameras. "The three cameras together function as the equivalent of a scout," said Bashar Rizk, who described their resolution as ranging from the periphery of human vision, to the center of the human eye, to a good pair of binoculars. "They're designed to check out the kind of environment we're going to be in. You'd like to be there yourself, but you're not able to do that, so we're sending these cameras instead."

The team won't have the luxury of collecting data and setting it aside for later analysis—they'll need to choose a safe and scientifically valuable site to scrape a sample within the 505 days of the prime mission. A major part of Beshore's job will be coordinating the various science teams to make sure that decision is made on schedule.

Its square solar arrays stuck out on either side like cumbersome wings, the spacecraft will sidle close to Bennu at the slow pace of a waltz and reach out a slender robotic arm to gather a sample. "That will be a terrifying moment," Lauretta said. "Mars has its seven minutes of terror, but we'll have four-and-a-half hours of waiting to get to the surface safely." Contact with the asteroid will last only a moment—team members refer to the sampler as "touch-and-go"—but the capsule's contents, secured for the long journey back, will keep scientists busy for years.

If all goes well, the spacecraft will have a thousand days to explore Bennu from orbit before heading home with its long-awaited cargo, at least 60 grams of a carbon-rich asteroid. "You have highly experienced people on this mission who pretty clearly will be retired by the time the sample gets back," Beshore said, "so you have to plan who will do those jobs in 2023. In many cases those people are students now, maybe even high school students."

Drake took care to gather a multigenerational team, with experienced mission hands alongside undergraduate students. His fascination with deep time—planets spun into shape by the raw forces of matter and energy—gave him the ability to look far forward as well. His commitment to fostering the next generation ensured that the mission he had envisioned and intended to see through would continue in his absence.

"We're explorers in the truest sense," Lauretta said. "We're going to a world that nobody has seen before up close. We don't know what we're going to find."

Kuiper's small laboratory formed out of a time of contention, rife with quarrels between the scientists who focused their telescopes on distant objects and those who dreamed of reaching nearby ones. Now, more than fifty years later, OSIRIS-REx begins to heal the breach. The mission requires a close partnership between spacecraft scientists and ground-based astronomers as it strives to shed light on the big questions.

Spacecraft missions irrevocably altered the way planetary science happens, as Kuiper foresaw all those years ago when America first turned its eyes to the Moon. Yet astronomy has returned to play a vital role as scientists reach out to ever-more-distant points of light. LPL's overarching mission—to explore worlds—demands an interweaving of disciplines equally vast, from the art of interpreting a ray of light to complex engineering of mechanical parts.

If anything connects the many people who make up the Lunar Lab, it must be a sense of shared space—not just the three buildings that LPL currently occupies on the University of Arizona campus, each rich in legacy, but also the shared space of the solar system, a ring of sunlight carved out in the emptiness, vast by our standards but infinitesimally small compared to the rest of the universe: humanity's common ground. In the desert night, they look up at the Arizona sky and imagine holding a piece of a star in their hands.

EPILOGUE

Worlds Beyond

I FINISHED COMPILING INTERVIEWS and writing most of this history in 2010, LPL's fiftieth anniversary year. The project gave me a fascinating glimpse into a half-century of planetary exploration and insight into how the preferences, personalities, and chance encounters of ordinary people shape what humans know about our place in the grandness of space and time.

I usually started my interviews with the question, "How did you become interested in planetary science?" Almost invariably they began, "It started when I was a kid." The same stories appeared again and again: a glimpse of Sputnik blinking in the night sky; watching Ranger and Apollo shoot for the Moon; building a telescope in the backyard. As generations passed, the inspirations became missions to Mercury and Mars, the stunning photography of the Voyagers, *Star Wars* and *Star Trek* sketching out possibilities—though the wonder of gazing up into a sea of stars never changed.

The stories resonated with me because they reflected my own love of space exploration, which arose first from the starry skies spangled over the desert where I grew up on the outskirts of Tucson, and those hours I spent awake as a child, lying on the hood of my mother's Toyota during meteor showers to count each sudden streak of light.

An equally important question was how scientists at LPL envisioned the future of their field. Fifty years after its creation, LPL had surpassed perhaps even Kuiper's expectations as a pioneer in teaching and research. It lived through—and, indeed, made happen—the transition from astronomy to planetary science, from points of light to worlds.

"LPL has been at the forefront of planetary science for fifty years, as the solar system's planets changed from objects in telescopes into places that had been visited by more and more sophisticated spacecraft," said Tim Swindle, LPL's current director. "At the time of LPL's founding, we knew of nine planets, a few dozen moons, and a few hundred asteroids. Now, we know of thousands of planets (most of them around stars other than the Sun), hundreds of moons, and hundreds of thousands of asteroids (a large fraction of them discovered by search programs at LPL). To remain at the forefront as the field continues to evolve, we will have to evolve."

Swindle notes that LPL is entering a new period of change. The faculty and staff that Kuiper and Sonett hired at the founding of the laboratory and department have largely moved on. Over the last fifty years, that remarkable cohort of people shaped the LPL enterprise, and in doing so shaped the field of planetary science. Now a new generation must take the lead in solar system exploration, and Tucson's prominence in the field assures that their interests and skills will contribute greatly to what humanity knows about its place in the universe.

"LPL is a place where you're encouraged to explore and try new things, and all the people here are always at the forefront of new techniques, new activities, new worlds," said Dolores Hill, the meteorite expert who works in the basement of the Kuiper Building where I transcribed interviews for four years. "It's really an exciting place to be."

The discovery of planets around other stars in the early nineties sparked a transition in the field of planetary science at least as profound as President Kennedy's announcement that America would reach the Moon—"a real sea-change," as William Hubbard described it. The possibilities of exoplanets have just begun to unfold as new techniques develop to measure their temperatures and surmise the characteristics of their surfaces.

"That expands our number of targets from nine (keeping Pluto as a planet) to hundreds and thousands of objects in nearby space to us," Dante Lauretta said. "Planetary science is going to blossom enormously because there's simply that many more planets out there that need to be studied."

Exoplanets create new opportunities to pursue the question that intrigues many planetary scientists as well as space aficionados everywhere: the origin of life and its distribution in the universe. OSIRIS-REx will carry home, cradled in its sample return capsule, a piece to this puzzle. The mission will help tell us if fragments of asteroids falling to Earth—the same streaks of light that once dazzled me as a child—planted the seeds to our origins. In this way, humanity's ventures into desolate space reflect back on blue-green Earth and its inhabitants.

For the mission team, the opportunity to start answering that question unites them with a sense of camaraderie and shared commitment unique to the field. "The people are tremendous, the conversations are exciting, and you're dealing with an aspect of science where you actually get to see our world against the backdrop of infinite space," said Bashar Rizk. "I want to be surprised. I want to go there and see something I've never even imagined before."

Perhaps the most profound discovery for understanding life on Earth would be the discovery of life elsewhere. "We know all these extremophiles that live in the weirdest places on Earth—three miles down, for pity's sake, eating rock," Alan Binder said. "Life on Earth is in every niche you can think of, eating anything you can think of, even the weirdest things in the world." Mars, our nearest neighbor, dominates the search for tenacious microbes clinging to soil grains in the ever-present frost line, but scientists have also turned their thoughts toward Europa, Titan, and "Goldilocks" exoplanets at exactly the right place in their orbits.

The search for life dominates public interest in exploring space, but it sparks useful science as well because it is an overriding theme that requires detailed knowledge of geology, chemistry, climate, and the physical history of a planet's changing surface. So far, Earth stands as the only data point in this search. There's no evidence yet to show if Mars or Europa harbor such secrets, beneath soil or under ice. Peter Smith, gazing out the window of his gemstone-strewn office in the Drake Building at the flowering palo verde trees, marveled at the capacity for life to adapt and flourish in lonely places, filling every niche and cranny.

"I don't know if the ice in the northern plains of Mars is where life took hold or not, but it's a good chance," Smith said during the hectic heydays of the Phoenix mission. "That's where water comes up to the surface. This is the one place you can get to it."

Scientists discovered water on Mars *again* in 2015—but this time it wasn't ancient or frozen. For several years, HiRISE had photographed dark streaks inching down the rocky slopes each spring and vanishing again in autumn. An undergraduate at the University of Arizona, Lujendra Ojha, first noticed those features and proposed they could be formed by water. Spectral data from the

Mars Reconnaissance Orbiter later added convincing evidence to that idea. The orbiter detected perchlorate salts—like those found by Phoenix and Curiosity—at the same locations, enough to lower the freezing point and allow water to sometimes melt and flow.

That discovery is a step toward understanding how life might take hold on Mars or other planets, what it needs to thrive, and what shape it takes to adapt to its surroundings. The tools that can give us these answers are constantly evolving, from the camera-heavy Rangers plunging into the Moon's crust to delicate Phoenix unfolding its solar panels like fans. Engineering has become part of planetary exploration more than anyone could have anticipated, and each new spacecraft mission brings its own challenges, whether it is balloons plummeting toward the surface of Titan or massive telescopes drifting just above Earth's atmosphere.

Spacecraft missions, the bread-and-butter of LPL's research, cannot yet reach the hundreds of exoplanets discovered in the Milky Way. The light of these distant worlds can only be studied in the telescope lens, yet they are geophysical objects with swirling atmospheres and strange surfaces to explain. Thus exoplanets bridge the longstanding rift between planetary science and astronomy, which was born in part from strong-willed personalities and in part from a true divergence in techniques.

"The foundation for a lot of planetary science starts on the ground," Steve Larson said. "The Kuiper Belt was discovered from the ground. That's a whole new realm of the solar system." Although CSS and Spacewatch are the last programs at LPL to utilize telescopes full-time, they make a quiet counterpoint to the cacophony of spacecraft missions. The mountain observatories to which Kuiper devoted so much of his life remain critical to the field, discovering objects that later become targets for closer views and tracing others too distant to approach.

"It's a joy to see each of these worlds turn into a place that you can get really familiar with," Nick Schneider said. "But the nature of science is changing for each of those worlds; the obligation of how much you have to explain is correspondingly higher. I do see us turning more and more worlds from complete unknowns to well-categorized places. The fact that we've found planets around other stars means that we've got a good set of fuzzy points of light that we have yet to explain."

Closer to home, spacecraft will undoubtedly give us second and third looks at the solar system, from Io's sulfurous volcanoes to the geysers on Enceladus.

New Horizons, launched in 2006 on a nine-year journey, made the first close approach to Pluto, thus rounding off the Grand Tour left unfinished by the Voyagers and beginning the exploration of Kuiper Belt, the "third zone" of the solar system. Even the space between planets and the diffuse medium that exists beyond the influence of our Sun hold wonders we have yet to understand.

Since the excitement of the Apollo era, the Moon has once again fallen out of fashion. But for many of LPL's original members, our nearest companion still exerts a mysterious pull. The lab's early students, now leaders in the field, harbor visions of returning, this time in person, to the dusty surface they examined so closely. "My dream all along has been that we would have people working and living on these other places," Chuck Wood said. "I think we know now it's a lot harder than we used to think it might be. But I still think we should do it."

Underlying these visions of the future is the knowledge that we must stand at a distance to look back and truly understand the planet we call home. Dispassionately, we can marvel at Venus's broiling atmosphere or wonder how water-carved Mars became so dry. It is with slightly less equanimity that scientists study Earth's processes and patterns. Those other worlds are mirrors reflecting back at us a remarkable picture of the only habitable planet we know.

For all its ambition, planetary science remains a small field. It was shaped in many ways by chance—Sputnik blinking into orbit, the rigid traditions of astronomical institutions, an unmapped and disregarded Moon, Kuiper's ardent personality. The beauty of the planets—wandering stars that revealed themselves as unique and marvelous worlds—played a fundamental role, just as it did in inspiring my early love of space exploration. Kuiper came to Tucson, after all, because of its vast heavens, unhindered by cloud or city light.

I hope, even as I write this, that kids are looking up at the sky all over the world and wondering.

ACKNOWLEDGMENTS

THIS BOOK BELONGS to many people. First and foremost, it belongs to Mike Drake, who had the vision and foresight to imagine that the memories of his faculty and staff were important enough to capture in print. I am more grateful than I can say for his kindness and guidance.

Amy Phillips played an important role in formulating the concept of this project. Special thanks go to Ewen Whitaker, Bob Strom, Bill Hartmann, Dolores Hill, and Ken Domanik for their enthusiasm and support. The Arizona Space Grant Consortium, expertly managed by Susan Brew, funded the first year of the original oral history project.

Yisrael Spinoza rose admirably to the challenge of creating the beautiful website that accompanies this material. Maria Schuchardt of the Space Imagery Center devoted many hours to the project, as well as office space and encouragement. Historical photographs were provided by Maria Schuchardt, Ewen Whitaker, Dale Cruikshank, John and Jane Spencer, Bill Hartmann, and the photo archivists at JPL and NASA Ames Research Center, Julie Cooper and Lynn Albaugh.

Pat Roemer made an enormous contribution by way of many long conversations over early drafts. I am also grateful to Laurel Wilkening, Melanie Lenart, Michael Dahlstrom, and anonymous reviewers for making valuable comments on all or part of the manuscript. Lori Stiles and Alan Fischer deserve special mention for their outstanding science journalism.

I spent many an evening in the basement of the Kuiper Building transcribing interviews. Thanks to the members of the Microprobe Lab for their cheerful companionship.

To my colleagues on the Phoenix team, especially Sanlyn Buxner, Rob Bovill, Kenny Fine, Jacob Egan, Carla Bitter, and Sara Hammond: thank you for welcoming me. Pat Woida first introduced me to Phoenix when the lander was still in the development stage in 2004. I have fond memories of long hours spent in the clean room with the Engineering Model of the Robotic Arm Camera and Enya playing in the background. I am still grateful, these many years later, to Megan McEvoy and Anne Struble for lending me the use of a deep freezer, centrifuge, and spectrometer in 2004 for an experiment that involved simulating Martian conditions.

I would not harbor such a deep love of science without the excellent work of the volunteers and board members of the Southern Arizona Regional Science and Engineering Fair (SARSEF). My gratitude in particular to the late Jack Johnson, a larger-than-life figure in my memories from a very young age, and to the people who are continuing his work: Paula Johnson, Kathleen Bethel, Pat Woida, Liz Baker, and many others—you know who you are!

Tom Wilson, my undergraduate advisor, has been a steadfast guide since I first met him at the science fair many years ago. John Keller and the other excellent folks I met at TOPS (Towards Other Planetary Systems) on Big Island, Hawaii, deepened my joy in the night sky.

It takes great courage to become a character in someone else's book. I am indebted to the faculty, staff, and students of LPL, both past and present, who agreed to share their experiences on and off the record. Often during the course of these interviews I felt almost superfluous as a writer and fortunate to be on the receiving end of so many wonderfully told stories. I have relied on the memories of numerous people to write this book. Any omissions or errors are my own. My apologies to those people I have left out of the manuscript or mentioned only briefly. I hope I have done justice to LPL's story, and I know many more tales of its coming-of-age can be told.

The following people gave recorded interviews during the original oral history project: Victor Baker, Alan Binder, Bill Bottke, Bill Boynton, Lyle Broadfoot, Robert Brown, Humberto Campins, Guy Consolmagno, George Coyne, Dale Cruikshank, Lyn Doose, Mike Drake, Paul Geissler, Joe Giacalone, Rick Greenberg, Bill Hartmann, Floyd Herbert, Dolores Hill, Jay Holberg, Lonnie Hood, Bill Hubbard, Don Hunten, Randy Jokipii, John Keller, Jozsef Kota,

Lynn Lane, Hal Larson, Steve Larson, Dante Lauretta, Larry Lebofsky, Gene Levy, John Lewis, Jonathan Lunine, Renu Malhotra, Don McCarthy, Alfred McEwen, Bob McMillan, Jay Melosh, Bill Merline, Alexander Pavlov, George Rieke, Bill Sandel, Nick Schneider, Adam Showman, Mel Simmons, Peter Smith, Chuck Sonett, John Spencer, Bob Strom, Tim Swindle, Mark Sykes, Spencer Titley, Martin Tomasko, Ewen Whitaker, Laurel Wilkening, and Chuck Wood.

Various people answered my questions through informal conversations, including Brad Castalia, Linda Hickcox, Arnold Davidson, Tara Bode, and Mary Guerrieri. Others gave interviews in the manuscript's final stages, including Dante Lauretta, Tim Swindle, Anna Spitz, Ed Beshore, Toby Owen, Alan Hildebrand, Carl Hergenrother, Uwe Fink, Lonnie Hood, Susan Brew, Humberto Campins, and Bashar Rizk. Don McCarthy kindly answered my questions about the 61-inch telescope and Steve Larson about the Catalina Sky Survey.

Let this book stand as a tribute to the fine people who have passed away since I first began writing it, among them, Mike Drake, Tom Gehrels, Chuck Sonett, Floyd Herbert, Don Hunten, Frank Low, Mel Simmons, Godfrey Sill, and Beryl Whitaker.

Heartfelt thanks goes to my editors at the University of Arizona Press—Kathryn Conrad, Allyson Carter, and Scott De Herrera—for believing in the project.

Finally, I am grateful to my family for reading drafts and urging me to publish—Mom, Dad, Kasondra, and Jessica, you are a continual inspiration. And I owe much to my husband for the incalculable gift of time to write and for his understanding on the evenings and weekends when I seemed to spend more time in outer space than on Earth. Thank you, Chris, for stargazing with me.

NOTES

All quotes were drawn from personal communications with the author, with the exceptions noted below.

PART ONE: ARRIVAL

16 *"distinguished-looking optical astronomer"* Frank J. Low, G. H. Rieke, and R. D. Gehrz, "The Beginning of Modern Infrared Astronomy," *Annual Review of Astronomy and Astrophysics* 45 (2007): 43–75.

17 *"so highly automated that no pencil is ever used"* Gerard P. Kuiper, "The Lunar and Planetary Laboratory II," *Sky & Telescope* XXVII, no. 2 (February 1964): 88–92.

20 *"While our problems parallel"* Gerard P. Kuiper, "The Lunar and Planetary Laboratory I," *Sky & Telescope* XXVII, no. 1 (January 1964): 4–7.

21 *"I used to cry about that seeing"* O. Richard Norton, "Master Optician, Master Observer," *Sky & Telescope* 89, no. 5 (May 1995): 81.

21 *"probably the best site in the world"* Ibid.

24 *"as reminders to the occupants"* Tom Gehrels, *On the Glassy Sea: An Astronomer's Journey* (New York: American Institute of Physics, 1988).

PART TWO: TARGET MOON

32   *"beginning of the age of instant science"*   Ewen A. Whitaker, *The University of Arizona's Lunar and Planetary Laboratory: Its Founding and Early Years* (Tucson: University of Arizona Printing-Reproductions Department, 1985).

32   *"This is a great day for science"*   R. Cargill Hall, *Lunar Impact: A History of the Project Ranger* (Washington, D.C.: NASA History Office, 1977).

35   *"the finest lunar atlas ever produced"*   Whitaker, *The University of Arizona's Lunar and Planetary Laboratory.*

39   *"a compromise between cost and precision"*   Harold L. Johnson, "The Design of Low Cost Photometric Telescopes," *Communications of the Lunar and Planetary Laboratory* 7, no. 111 (1967).

50   *"That is not an uninteresting field"*   Alan S. Stern, "Forging a New Solar System," *Astronomy* 27, no. 3 (1999): 40–46.

51   *"spirit of discovery"*   University of Arizona Special Collections, Gerard P. Kuiper Papers.

51   *"I see no alternative"*   Ibid.

52   *"elegance and beauty"*   Derek W. G. Sears, "Report: Oral Histories in Meteoritics and Planetary Science—XX: Dale Cruikshank," *Meteoritics & Planetary Science* 48, no. 4 (2013): 1–12.

PART THREE: TEACHING

56   *"The astronomy department at the present time"*   University of Arizona Special Collections, Gerard P. Kuiper Papers.

66   *"cosmetically appealing"*   L. Ralph Baker, "Pioneer 11 Saturn Display System," *The Society of Motion Picture and Television Engineers Motion Imaging Journal* 89, no. 8 (August 1, 1980): 557–60.

73   *"the most important ice in the solar system"*   Uwe Fink and Godfrey T. Sill, "The Infrared Spectral Properties of Frozen Volatiles," in *Comets*, edited by Laurel Wilkening, 164–202 (Tucson: University of Arizona Press, 1982).

74   *"These were the days of the US space program"*   Gehrels, *On the Glassy Sea.*

76   *"I came away from the Saturn operations"*   Ibid.

PART FOUR: VOYAGES

81   *"the nation's tour guide"*   Bradford Smith, *People* (December 28, 1981).

81   *"the existing atmospheric circulation models"*   Quoted in George E. Webb, *Science in the American Southwest: A Topical History* (Tucson: University of Arizona Press, 2002).

96   *"turned the Earth's surface into a living hell"*   Alan R. Hildebrand, "The Cretaceous/Tertiary Boundary Impact (or the Dinosaurs Didn't Have a Chance)," *Journal of the Royal Astronomical Society of Canada* 87, no. 2 (1993): 77–118.

98   *"What's a pixel?"*   David H. Levy, "The New Age of CCD Observing," in *Star Trails: 50 Favorite Columns from* Sky & Telescope (Cambridge, MA: Sky Publishing, 2007), 31–34.

100   *"We should not ignore the possibility"*   K. Serkowksi, "Should We Search for Planets around Other Stars?" *Astronomy Quarterly* 1, no. 5 (1997), quoted in Gehrels, *On the Glassy Sea*, 195.

PART FIVE: RETURNING

111   *"the hypothesis of last resort"*   Carl Sagan et al., "A Search for Life on Earth from the Galileo Spacecraft," *Nature* 365 (October 21, 1993): 715–21.

114   *"This looks like a squashed comet!"*   David H. Levy, *Impact Jupiter: The Crash of Comet Shoemaker-Levy 9* (New York: Plenum, 1995).

116   *"It's simply a beautiful structure"*   David Perlman, "Baffling Seismic Fault Seen in Icy Crust of Jovian Moon," *San Francisco Chronicle* (December 8, 1998).

117   *"It has beauty not because we are one with it"*   Quoted in Neil Miller, *Kartchner Caverns: How Two Cavers Discovered and Saved One of the Wonders of the Natural World* (Tucson: University of Arizona Press, 2008).

FURTHER READING

For those interested in learning more about the history of the Lunar and Planetary Lab and the development of the field of planetary science, refer to the online companion to this book at www.lpl.arizona.edu/history, a website that gathers excerpts from more than fifty interviews compiled in 2006–10. Also refer to Ewen Whitaker's history *The Lunar and Planetary Laboratory: Its Founding and Early Years* (University of Arizona Printing-Reproductions Department, 1985) and the other references listed below.

BOOKS

Corfield, Richard. *Lives of the Planets: A Natural History of the Solar System*. New York: Basic Books, 2007.

Doel, Ronald E. *Solar System Astronomy in America: Communities, Patronage, and Interdisciplinary Research, 1920–1960*. Cambridge, MA: Cambridge University Press, 1996.

Frankel, Charles. *The End of the Dinosaurs: Chicxulub Crater and Mass Extinctions*. Cambridge, MA: Cambridge University Press, 1999.

Gehrels, Tom. *On the Glassy Sea: An Astronomer's Journey*. New York: American Institute of Physics, 1988.

Greenberg, Richard. *Unmasking Europa: The Search for Life on Jupiter's Ocean Moon*. New York: Copernicus Books, 2008.

Hartmann, William K., and Ron Miller. *The History of Earth: An Illustrated Chronicle of an Evolving Planet.* New York: Workman, 1991.

Hensel, Suzanne. *Look to the Mountains: An In-Depth Look into the Lives and Times of the People Who Shaped the History of the Catalina Mountains.* Tucson, AZ: Mt. Lemmon Woman's Club, 2003.

Hubbard, Scott. *Exploring Mars: Chronicles from a Decade of Discovery.* Tucson: University of Arizona Press, 2011.

Levy, David H. *Impact Jupiter: The Crash of Comet Shoemaker-Levy 9.* New York: Plenum, 1995.

———. *The Man Who Sold the Milky Way: A Biography of Bart Bok.* Tucson: University of Arizona Press, 1993.

Mackenzie, Dana. *The Big Splat, or How Our Moon Came to Be.* Hoboken, NJ: Wiley, 2003.

McCray, W. Patrick. *Giant Telescopes: Astronomical Ambition and the Promise of Technology.* Cambridge, MA: Harvard University Press, 2004.

Osterbrock, Donald E. *Yerkes Observatory, 1892–1950: The Birth, Near Death, and Resurrection of a Scientific Research Institution.* Chicago: University of Chicago Press, 1997.

Rowan-Robinson, Michael. *Night Vision: Exploring the Infrared Universe.* Cambridge, MA: Cambridge University Press, 2013.

Sheehan, William. *Worlds in the Sky: Planetary Discovery from Earliest Times through Voyager and Magellan.* Tucson: University of Arizona Press, 2013.

Tatarewicz, Joseph N. *Space Technology and Planetary Astronomy.* Bloomington: Indiana University Press, 1990.

Webb, George E. *Science in the American Southwest: A Topical History.* Tucson: University of Arizona Press, 2002.

Whitaker, Ewen A. *Mapping and Naming the Moon: A History of Lunar Cartography and Nomenclature.* Cambridge, MA: Cambridge University Press, 1999.

Wilhelms, Don E. *To a Rocky Moon: A Geologist's History of Lunar Exploration.* Tucson: University of Arizona Press, 1993.

HISTORIES AND OFFICIAL REPORTS

Augason, Gordon C., and Hyron Spinrad. *Infrared Astronomy Review for the Astronomy Subcommittee of the National Aeronautics and Space Administration.* Washington, DC: NASA Ames Research Center, 1965.

Blenman, Charles. *Final Report: Imaging Photopolarimeter Experiment in Modes 3 and 4.* Washington, DC: NASA Ames Research Center, 1981.

Byres, Bruce K. *Destination Moon: A History of the Lunar Orbiter Program.* Washington, DC: NASA History Office, 1977.

Cortright, Edgar M., ed. *Apollo Expeditions to the Moon.* Washington, DC: NASA History Office, 1975.

Dolci, Wendy Whiting. *Milestones in Airborne Astronomy: From the 1920's to the Present.* Reston, VA: American Institute of Aeronautics and Astronautics, 1997.

Fimmel, Richard O., William Swindell, and Eric Burgess. *Pioneer Odyssey.* Washington, DC: NASA History Office, 1977.

Hall, R. Cargill. *Lunar Impact: A History of the Project Ranger.* Washington, DC: NASA History Office, 1977.

Hartman, Edwin, P. *Adventures in Research: A History of Ames Research Center, 1940–1965.* Washington DC: NASA History Office, 1970.

Low, Frank. "Airborne Infrared Astronomy: The Early Days." In *NASA Conference Publication #2353 Airborne Astronomy Symposium—Proceedings of a Meeting Sponsored by NASA and the ASP.* Mountain View, CA: NASA Ames Research Center, 1984.

Low, Frank J., G. H. Rieke, and R. D. Gehrz. "The Beginning of Modern Infrared Astronomy." *The Annual Review of Astronomy and Astrophysics* 45 (2007): 43–75.

Meltzer, Michael. *Mission to Jupiter: A History of the Galileo Project.* Washington, DC: NASA History Division, 2007.

O'Donnell, Franklin. *Explorer I.* Pasadena, CA: California Institute of Technology, 2007.

Whitaker, Ewen A. *The University of Arizona's Lunar and Planetary Laboratory: Its Founding and Early Years.* Tucson: University of Arizona Printing-Reproductions Department, 1985.

SCIENTIFIC ARTICLES

Baker, A. L., et al. "The Imaging Photopolarimeter Experiment on Pioneer 11." *Science* 188, no. 4187 (May 2, 1975): 468–72.

Boynton, W. V., et al. "Evidence for Calcium Carbonate at the Mars Phoenix Landing Site." *Science* 325, no. 61 (2009): 61–64.

Coyne, G. V., and T. Gehrels. "Polarimetry from High Altitude Balloons." *Communications of the Lunar and Planetary Laboratory* 7, no. 108 (February 8, 1968).

Drake, Michael J. "Accretion and Primary Differentiation of the Earth: A Personal Journey." *Geochimica et Cosmochimica Acta* 64, no. 14 (2000): 2363–69.

Fink, Uwe, and Godfrey T. Sill. "The Infrared Spectral Properties of Frozen Volatiles." In *Comets*, edited by Laurel Wilkening, 164–202. Tucson: University of Arizona Press, 1982.

Gehrels, Tom. "Ultraviolet Polarimetry Using High Altitude Balloons." *Applied Optics* 6, no. 2 (1967): 231–33.

Gehrels, Tom, et al. "Imaging Photopolarimeter on Pioneer Saturn." *Science* 207 (January 25, 1980): 434–39.

Geissler, Paul, W. Reid Thompson, Richard Greenberg, Jeff Moersch, Alfred McEwen, and Carl Sagan. "Galileo Multispectral Imaging of Earth." *Journal of Geophysical Research* 100, no. E8 (August 25, 1995): 16895–906.

Hartmann, William K., and Donald R. Davis. "Satellite-Sized Planetesimals and Lunar Origin." *Icarus* 24 (1975): 504–15.

Hecht, M. H., et al. "Detection of Perchlorate and the Soluble Chemistry of Martian Soil at the Phoenix Lander Site." *Science* 325, no. 64 (2009): 64–67.

Hildebrand, Alan R., et al. "Chicxulub Crater: A Possible Cretaceous/Tertiary Boundary Impact Crater on the Yucatan Peninsula, Mexico." *Geology* 19 (September 1991): 867–71.

Hildebrand, Alan R. "The Cretaceous/Tertiary Boundary Impact (or the Dinosaurs Didn't Have a Chance). *Journal of the Royal Astronomical Society of Canada* 87, no. 2 (1993): 77–118.

Hobbs, B. A., L. L. Hood, F. Herbert, and C. P. Sonett. "An Upper Bound on the Radius of a Highly Electrically Conducting Lunar Core." *Journal of Geophysical Research* 88, no. S01 (November 10, 1983): B97–102.

Hood, Lonnie L. "The Enigma of Lunar Magnetism. *Eos* 62, no. 16 (April 21, 1981): 161–63.

Hunten, Donald M. "An Unplanned Career in Space Physics." *Journal of Geophysical Research* 101, no. A5 (May 1, 1996): 10567–76.

Irion, Robert. "Lunar Prospector Probes Moon's Core Mysteries." *Science* 281, no. 5382 (September 4, 1998): 1423–25.

Johnson, Harold L. "The Design of Low-Cost Photometric Telescopes." *Communications of the Lunar and Planetary Laboratory* 7, no. 111 (1967).

Kuiper, Gerard P. "The Lunar and Planetary Laboratory and Its Telescopes." *Communications of the Lunar and Planetary Laboratory* 9, no. 172 (1972): 199–245.

———. "The Lunar and Planetary Laboratory I." *Sky & Telescope* XXVII, no. 1 (1964): 4–7.

———. "The Lunar and Planetary Laboratory II." *Sky & Telescope* XXVII, no. 2 (1964): 88–92.

———. "Organization and Programs of the Laboratory." *Communications of the Lunar and Planetary Laboratory* 1, no. 1 (February 12, 1962): 1–20.

———. "Preliminary Determination of the Bearing Strength of the Floor of the Crater Alphonsus." *JPL Technical Report No. 32–800* (1980).

Kuiper, Gerard P., F. F. Forbes, and H. L. Johnson. "A Program of Astronomical Infrared Spectroscopy from Aircraft." *Communications of the Lunar and Planetary Laboratory* 6, no. 93 (June 30, 1967).

Ojha, Lujendra et al. "Spectral Evidence for Hydrated Salts in Recurring Slope Lineae on Mars." *Nature Geoscience*, advanced online publication (September 28, 2015).

Schulte, Peter, et al. "The Chicxulub Asteroid Impact and Mass Extinction at the Cretaceous-Paleogene Boundary." *Science* 327 (March 5, 2010): 1214–18.

Sill, Godfrey T. "Sulfuric Acid in the Venus Clouds." *Communications of the Lunar and Planetary Laboratory* 9, no. 171 (1972).

Smith, Peter, et al. "H_2O at the Phoenix Landing Site." *Science* 325, no. 58 (2009): 58–61.

Smith, Peter, and Lyn Doose. "Observing Saturn from Pioneer 11." *Sky & Telescope* 58 (1978): 405–8.

Spradley, L. H. "Lunar Globe Photography." *Communications of the Lunar and Planetary Laboratory* 1, no. 6 (January 26, 1962): 31–34.

Tomasko, M. G., et al. "Preliminary Results of the Solar Flux Radiometer Experiment aboard the Pioneer Venus Multiprobe Mission." *Science* 203 (February 23, 1979): 795–97.

Tomasko, M. G. et al. "The Descent Imager/Spectral Radiometer (DISR) Experiment on the Huygens Entry Probe of Titan." *Space Science Reviews* 104 (2002): 469–551.

Whitaker, Ewen A. "Surveyor 1 Location." *Communications of the Lunar and Planetary Laboratory* 6, no. 83 (1967): 49–50.

POPULAR PRESS

Cowen, Ron. "Liquid Acquisition: Two New Scenarios Ramp Up Debate over How Earth Got Its Water." *Science News* (January 15, 2011): 26–29.

Cruikshank, Dale P. "20th-Century Astronomer." *Sky & Telescope* 47, no. 3 (1974): 159–64.

Cruikshank, Dale P. "Gerard Peter Kuiper." *Biographical Memoirs* 62 (1993): 259–95.

De Vaucouleurs, Gérard H. "Harold Lester Johnson." *Biographical Memoirs* 67 (1995): 243–61.

"Eighteen from UA on Teams Studying Pioneer Mission." *Arizona Daily Star* (August 30, 1979).

Farnsworth, Aaron. "Jupiter's Moon: UA Scientist Finds Water and Life-Sustaining Nutrients on Europa." *University of Arizona Report on Research* (2002): 20–21.

Fischer, Alan. "Research in the Desert Paying Off in the Stars." *Tucson Citizen* (May 20, 2008).

Greenberg, Richard. "B. Randall Tufts." *Eos* 83, no. 41 (October 8, 2002): 460–61.

Haas, W. H. "In Memoriam: Alika K. Herring." *Journal of the Association of Lunar and Planetary Observers* 40, no. 1 (1998): 30–31.

Knetzger, Jack. "Probing the Ringed Planet." *Arizona Daily Wildcat* (August 28, 1979).

———. "Refining Images an Intricate Task." *Arizona Daily Wildcat* (August 28, 1979).

"Kuiper's Interest Closer to Home." *Arizona Daily Star* (August 1, 1964).

Maggio, Elizabeth. "Noted UA Astronomer Gerard P. Kuiper Dies." *Arizona Daily Star* (December 25, 1973).

———. "UA Scientists Provide Pioneer 11's 'Eyes' for Flyby this Week." *Lo Que Pasa* 27 (August 1979).

Norton, O. Richard. "Master Optician, Master Observer." *Sky & Telescope* 89, no. 5 (May 1995): 81.

Overbye, Dennis. "Frank J. Low, Who Helped Drive Field of Infrared Astronomy, Dies at 75." *New York Times* (June 20, 2009).

Owen, Tobias, and Carl Sagan. "Obituary: Planetary Astronomer, Gerard Peter Kuiper." *Mercury: The Journal of the Astronomical Society of the Pacific* 3, no. 2–3 (1974): 16–18, 37.

"Photos of Moon Analyzed by U of A's Dr. Kuiper." *Arizona Daily Star* (August 1, 1964).

Righter, Kevin, John Jones, David Mittlefehldt, Allan Treiman, and Nancy Chabot. "Michael J. Drake (1946–2011)." *Geochemical Society* (September 2011).

Ryman, Anne. "UA Scientist Leads Mission to Mars." *Arizona Republic* (July 29, 2007).

Sagan, Carl. "Obituary: Gerard Peter Kuiper." *Icarus* 22 (1974): 117–18.

Schilling, Govert. "Remembering Tom Gehrels (1925–2011)." *Sky & Telescope News* (July 12, 2011).

Scotti, James. "Obituary: Wieslaw Z. Wisniewski, 1931–1994." *Bulletin of the American Astronomical Society* 26, no. 4 (1994): 1611–12.

Smith, Grant E. "Spacecraft Will Send Photos of Saturn 960 Million Miles." *Arizona Republic* (August 19, 1979).

Stern, Alan S. "Forging a New Solar System." *Astronomy* 27, no. 3 (1999): 40–46.

Stiles, Lori. "Galileo Mission Arrives at Jupiter Today." *Lo Que Pasa* (December 7, 1995).

———. "Huygens Probe: Interplanetary Journey Reveals Secrets of Titan." *University of Arizona Report on Research* 21, no. 2 (2005).

———. "LPL's July 20 Open House Celebrates Lunar Missions, Birth of Lab." *UA News Services* (July 12, 1999).

———. "Lunar and Planetary Lab History Is Told by Those Who Made It. *UA News Services* (December 17, 2008).

———."UA Lunar and Planetary Lab Moves into Two Additional Buildings." *UA News Services* (May 20, 2004).

———. "UA Observers of Comet SL9 Impact." *UA News Services* (July 17, 1994).

———."UA's Cassini Scientists Ready for First Close Titan Flyby. *UA News Services* (October 25, 2004).

Stolte, Daniel. "Charles P. Sonett: The Legacy of a Pioneering Space Scientist." *UA News Service* (October 5, 2011).

Sorenson, Dan, and Eric Swedlund. "Was there Life on Mars?" *Arizona Daily Star* (July 29, 2007).

Swedlund, Eric. "AZ's Long Road to Mars." *Arizona Daily Star* (July 30, 2007).

———."Going Beyond." *Arizona Daily Star* (July 31, 2007).

"UA Dreams Stretch from Moon to Mars." *Arizona Daily Star* (July 12, 2009).

INDEX

ABOUT THE AUTHOR

Melissa L. Sevigny (née Lamberton) grew up in Tucson, Arizona, with a deep love of the geology, ecology, and clear desert skies of the Southwest. She holds bachelor's degrees in environmental science and creative writing from the University of Arizona and an MFA in creative writing and environment from Iowa State University. She was the education and public outreach coordinator of NASA's Phoenix Mars Scout Mission during ground operations, and has worked as a science communicator in the fields of planetary science, water resources, and sustainable agriculture. She is currently the science and technology reporter for KNAU (Arizona Public Radio) in Flagstaff. Minor Planet (15624) Lamberton is named in her honor.